Lecture Notes in Physics

Volume 1036

The series Lecture Notes in Physics (LNP), founded in 1969, reports new developments in physics research and teaching - quickly and informally, but with a high quality and the explicit aim to summarize and communicate current knowledge in an accessible way. Books published in this series are conceived as bridging material between advanced graduate textbooks and the forefront of research and to serve three purposes:

- to be a compact and modern up-to-date source of reference on a well-defined topic;
- to serve as an accessible introduction to the field to postgraduate students and non-specialist researchers from related areas;
- to be a source of advanced teaching material for specialized seminars, courses and schools.

Both monographs and multi-author volumes will be considered for publication. Edited volumes should however consist of a very limited number of contributions only. Proceedings will not be considered for LNP.

Volumes published in LNP are disseminated both in print and in electronic formats, the electronic archive being available at springerlink.com. The series content is indexed, abstracted and referenced by many abstracting and information services, bibliographic networks, subscription agencies, library networks, and consortia.

Proposals should be sent to a member of the Editorial Board, or directly to the responsible editor at Springer:

Dr Lisa Scalone
lisa.scalone@springernature.com

Sumati Surya

The Causal Set Approach to Quantum Gravity

An Introduction

Springer

Sumati Surya
Raman Research Institute
Bengaluru, Karnataka, India

ISSN 0075-8450 ISSN 1616-6361 (electronic)
Lecture Notes in Physics
ISBN 978-3-031-84419-5 ISBN 978-3-031-84420-1 (eBook)
https://doi.org/10.1007/978-3-031-84420-1

This Springer imprint is published by the registered company Springer Nature Switzerland AG
The registered company address is: Gewerbestrasse 11, 6330 Cham, Switzerland

To Amma and Appa who gave me the freedom to question.

Preface

This book is intended for current practitioners in quantum gravity, as well beginners to research in the field. It aims to give a broad overview of the contours of the causal set approach, starting from the main hypothesis, through some well-established results, arriving all the way to the frontiers of our understanding. It presents some of the key results in causal set theory, with enough detail to make it possible for the interested reader to begin working in the subject. Scattered throughout the book are exercises that will give the reader simple practice with causal sets.

Chapter 1 offers a non-technical overview of Causal Set Theory (CST) within the context of quantum gravity, while Chap. 2 introduces the causal set hypothesis, foundational theorems in Lorentzian geometry, and the discrete-continuum approximation that supports CST's core ideas. Chapter 3 explores how continuum geometry and topology are approximated by random causal sets, showcasing CST's ability to encode continuum spacetime structures in a discrete substratum. Chapter 4 examines quantum scalar fields on causal sets, discussing the potential effects of discreteness on quantum fields and introducing a causal set based calculation of entanglement entropy. CST dynamics, covered in Chap. 5, leverages nonlocality in the non-perturbative CST path sum, to prove that the most dominant non-spacetimelike contributions are suppressed. This chapter also elaborates on the Lorentzian state-sum models and the more fundamental sequential growth models. The final Chap. 6 is a brief discussion of CST's phenomenological implications, starting with Sorkin's surprisingly accurate 1987 prediction for the cosmological constant, and the discreteness induced particle "swerves", ending with a short list of open questions.

Bengaluru, India
August 2025

Sumati Surya

No one can teach you aught but that which lies half asleep in the dawning of your knowledge.

—Khalil Gibran

Acknowledgements The quote above captures how the relativist inside us might respond to the ideas behind the causal set approach to quantising gravity.

This book is a result of the work and dedication of those who have contributed to causal set theory. Rafael Sorkin's deep insight into Lorentzian geometry and quantum foundations has helped forge the two into a coherent theory of quantum gravity. I am forever indebted to him for opening the doors to a whole new way of thinking of physics, and for his generosity in sharing his ideas. Fay Dowker has been a long term collaborator and friend without whose encouragement I may never have begun the daring foray into causal set waters. Her ability to concretise vague thoughts into actionable mathematical ideas has been invaluable, and my understanding has been enriched by countless discussions with her. I am also very lucky to have worked with many others on causal set theory who I list here in alphabetical order: Anish Bhattacharya, Graham Brightwell, Steve Carlip, Will Cunningham, Astrid Eichhorn, L. Glaser, Joe Henson, Steven Johnston, Ian Jubb, Seth Major, Abhishek Mathur, Denjoe O'Connor, Abhishek Ravishankar, Kasia Rejzner, David Rideout, Alan Daniel Santhosh, Anup Anand Singh, Ritesh Srivastava, Fleur Versteegen, Nomaan X., Yasaman Yazdi and Stav Zalel. Together, we have scratched our heads and argued our way through thick forests of ideas, making progress when possible, leaving some ideas to slowly simmer. This community has given each of us enough room to be creative, while at the same time providing ample stimulation, thanks in large measure to the energies and insights of Rafael Sorkin and Fay Dowker.

I am deeply grateful to the Raman Research Institute for providing a calm, supportive environment over the years, conducive to work in quantum gravity. Work like ours would not be possible without several kinds of help. I thank the administration at RRI, especially Manju, Gayathri and Mahadeva for their kind support over the years.

A special thanks to Nomaan X for regenerating plots from unused data and for Christoph Minz for his causal set graphics package. I also thank Abhishek Mathur for his patient reading of parts of the text. During the last stages of writing this book I was also partly supported by the Blaumann Foundation Grant and the SERB MATRICS Grant Number MTR/2023/000831.

I thank my editors for their patience (and final impatience) at my continual missing of deadlines. I would have liked to write this book over a decade, slowly and with care, but as they say, time waits for no one.

Finally, I am delighted to thank my family: Sachin, Mythili, Yasmin, Amma, Appa and Arjun, for their love and patience as I slogged through the writing of this book. Thanks in particular to Yasmin who helped me type in changes to the manuscript when I was nursing an injury. It is my hope that our daughters will continue to carry with them a love of mathematics and the natural world as they make their way in this increasing complex world.

Qn. Do you think you can travel back in time?

Ans. Not in Reality. Only in your imagination.

Conversation with Yasmin, aged 12

Bengaluru, India
August 2025

Sumati Surya

Competing Interests The author has no competing interests to declare that are relevant to the content of this manuscript.

Contents

Acronyms

BDG	Benincasa-Dowker-Glaser
CHK	Caratheodory-Hahn-Kluvanek
CNN	Convex Normal Neighbourhood
CSG	Classical Sequential Growth
ℂSG	Complex Sequential Growth
CST	Causal Set Theory
EE	Entanglement Entropy
GNN	Gaussian Normal Neighbourhood
GPPP	General Poisson Point Process
HH	Hartle-Hawking
HKMMKP	Hawking-King-McCarthy-Malament-Kronheimer-Penrose
KCH	Kolmogorov-Caratheodary-Hahn
KR	Kleitmann-Rothschild
MCMC	Markov Chain Monte Carlo
PJ	Pauli-Jordan
QSG	Quantum Sequential Growth
RNN	Riemann Normal Neighbourhood
SJ	Sorkin-Johnston
SOH	Sum-Over-Histories
SSEE	Sorkin's Spacetime Entanglement Entropy

Notations

General Notation

$\mathbb{N}$	Natural Numbers
$\mathbb{R}$	Real Line
$\mathbb{C}$	Complex Plane
$\lvert A\rvert$	Cardinality of the set A
A^c	Complement of a subset $A \subset B$ in the set B
A	Closure of a set A

Continuum Notation

$\prec$	Irreflexive causality relation $x \prec y$, $x \nprec x$
$\preceq$	Reflexive causality relation, with $x \preceq y \preceq x \Rightarrow x = y$
$\prec\!\!\prec$	Irreflexive chronology relation $x \prec\!\!\prec y$
$\rightarrow$	Horismotic or light-like relation $x \rightarrow y$, reflexive
$\parallel$	Spacelike relation
$I^{\pm}(p)$	Chronological future and past, respectively, of an event p
$J^{\pm}(p)$	Causal future and past, respectively, of an event p
$I(p, q)$	$\equiv \{r \mid p \prec\!\!\prec r \prec\!\!\prec q\}$ is an **Alexandrov interval**
$\mathrm{vol}_g(X)$	Spacetime volume of the region X in a spacetime (M, g)
$\mathrm{vol}(p, q)$	$\equiv \mathrm{vol}(I(p, q))$, spacetime volume of $I(p, q)$
$\mathbb{M}^d$	d-dimensional Minkowski spacetime
η_{ab}	Minkowski metric
$\mathbb{D}^d$	$\equiv I(p, q) \subset \mathbb{M}^d$ for some $p, q \in \mathbb{M}^d$
$\tau(p, q)$	Proper time from p to q
$\mathrm{vol}(\mathbb{D}^d)$	$\equiv \zeta_d \tau(p, q)^d$
ζ_d	$\equiv S_{d-2} \frac{2^{1-d}}{d(d-1)}$

S_{d-2}	$\equiv \frac{2\pi^{(d-1)/2}}{\Gamma\left(\frac{d-1}{2}\right)}$, the volume of the round $(d-2)$-dimensional unit sphere $\mathbb{S}^{d-2}$
$\mathbf{s}_d$	$\equiv S_{d-2}\frac{1}{d(d-1)}$

Causal Set Notation

$\prec$	Irreflexive order relation
$\|\|$	Incomparable or spacelike: $e \,\|\|\, e' \Rightarrow e \not\prec e' \;\&\; e' \not\prec e$
C	Causal set with an irreflexive order relation $\prec$
Fut(e)	$\equiv \{e' \in C \mid e \prec e'\}$, **Future of** e
Past(e)	$\equiv \{e' \in C \mid e' \prec e\}$, **Past of** e
IFut(e)	$\equiv \{e' \in C \mid e \preceq e'\}$, **Inclusive future of** e
IPast(e)	$\equiv \{e' \in C \mid e' \preceq e\}$, **Inclusive past of** e
(e, e')	$\equiv \{e'' \in C \mid e \prec e'' \prec e'\} = \mathrm{Fut}(e) \cap \mathrm{Past}(e')$, **Order interval**
$[e, e']$	$\equiv \{e'' \in C \mid e \preceq e'' \preceq e'\} = \mathrm{IFut}(e) \cap \mathrm{IPast}(e')$, **Closed order interval**
$M_C(e, e')$	**Causal Matrix element**
$\mathfrak{C}_\rho(M, g)$	Ensemble of causal sets generated from a Poisson sprinkling into (M, g) at density ρ
A^{cc}	Causal complement of the subcausal set $A \subset C$ in C

Overview 1

To admit structures which can be very different from a manifold. The possibility arises, for example, of a locally countable or discrete event-space equipped with causal relations macroscopically similar to those of a space-time continuum.
– Kronheimer and Penrose [1].

Abstract

This chapter gives a broad and largely non-technical introduction to causal set theory (CST), placing it within the larger context of our search for a quantum theory of gravity. It provides a detailed overview of the subsequent chapters of the book, which are ordered to maintain a pedagogical flow. The reader may find this broad perspective useful to come back to from time to time, as they traverse this book.

To a practitioner of quantum gravity it is obvious that the "problem of quantum gravity" is *very* hard, and that despite several decades of concerted effort, it remains largely unsolved. Time has spawned several approaches to this problem, some of which involve hundreds of workers. While none of these approaches can be deemed fully successful, each affords us an insight into quantum gravity—what it could be, and perhaps what it is not.

This book is an exposition of the approach known as causal set theory (CST) [2]. In this approach the spacetime continuum is replaced by discrete substructures which are locally finite partially ordered sets or **posets**. Lorentzian geometry is the rich mathematical soil in which CST germinated, and to which it is firmly tethered. This lends a unique and deeply Lorentzian perspective on quantum gravity which, over the years, has enriched the field of quantum gravity research at large.

S. Surya, *The Causal Set Approach to Quantum Gravity*, Lecture Notes in Physics 1036,
https://doi.org/10.1007/978-3-031-84420-1_1

CST gives centrality to two principles: causality and spacetime discreteness. Causality is widely considered to be an important principle of physics, both in the classical and the quantum domain. Spacetime discreteness, on the other hand, is often motivated by semi-classical or quantum considerations, for example, the need for UV finiteness. It implies that continuum spacetime geometry is "emergent" and not fundamental. CST combines these two principles in a manner that is deeply inspired by the central role played by the spacetime causal structure in Lorentzian geometry.

The principle of causality is tied to the Lorentzian character of spacetime geometry. By a spacetime or a Lorentzian geometry we mean specifically the signature $(-, +, + \dots +)$, which has a single time direction. This in turn implies that the tangent space at every spacetime event is split up into disjoint sets of past, future and spacelike vectors. The disjointness of the past from the future is a critical topological property, absent not only in Riemannian spaces but also in spacetimes with other signatures, i.e., those with more than one time direction. The events in a spacetime (M, g) with no closed causal curves (causal spacetimes) are **partially ordered** by the causality relation $\prec$, which means there is an underlying **causal structure poset** $(M, \prec)$. CST uses a discretisation of the causal structure poset to motivate its order theoretic ontology.

The principles of causality and discreteness are not radical, and while they are shared by some of the other approaches to quantum gravity, they differ from CST in their implementation. For example, causality is a fundamental principle in causal dynamical triangulations [3], and in causal spin-foam models [4,5]. However, even though these approaches are formulated using discrete structures, as in lattice gauge theory, the physics is thought to emerge only in the continuum limit. On the other hand, in approaches like non-commutative geometry, discreteness of the phase space is taken to be fundamental, but the formulation does not emphasise the Lorentzian structure [6]. It can be argued that the twin principles of causality and fundamental spacetime discreteness, combined with an explicitly constructed continuum approximation, lead inevitably to CST [7].

Although CST was crystallised in the 1987 paper by Bombelli, Lee, Meyer and Sorkin [2], its origins lie in the slow unveiling of the role of causal structure in Lorentzian geometry. The earliest realisation of the role of order in spacetime goes back to Robb's 1914 analysis of special relativity, and his construction of the "conical order" on Minkowski spacetime [8]. Many years later, Alexandrov and Ovchinnikova [9,10], and independently Zeeman [11], showed that the group of causal automorphisms in Minkowski spacetime of dimension $d > 2$ is equivalent to the group of Poincare transformations plus dilatations. Subsequently, Kronheimer and Penrose reimagined spacetime using an axiomatisation of the causal structure and proposed this as a starting point for spacetime quantisation. A series of theorems by Hawking [12], Hawking, King, McCarthy [13] and Malament [14], aided by a result from Kronheimer and Penrose [1], then collectively showed that for causal spacetimes which satisfy certain weak causality properties, the causal structure poset encodes the complete conformal geometry of the spacetime. Thus, the only remaining degree of freedom *not* contained in the causal structure poset of such a spacetime is the conformal factor, or equivalently, the local volume element. This Hawking-King-

McCarthy-Malament-Kronheimer-Penrose (HKMMKP) theorem thus give us a way to reimagine spacetime geometry as the causal structure "plus" the volume element. As Finkelstein quipped, the causal structure poset is "9/10th" of the spacetime metric in four dimensions [15]!

In any approach to quantum gravity one must start with a choice of the fundamental degrees of freedom that are to be quantised. In loop quantum gravity it is the phase space associated with the $SU(2)$ connection [16], in causal dynamical triangulations it is the quantum geometry obtained from a causal simplicial decomposition [3], while in string theory, gravity emerges from a theory of vibrating strings and branes [17]. The HKMMKP theorem gives us a quintessentially Lorentzian candidate for quantisation, namely, the causal structure poset.

The idea of quantising or discretising the causal structure was suggested by several authors, prominently by Kronheimer and Penrose [1], Finkelstein [15] and by Myrheim, in an ignored CERN preprint [18]. The causal set approach to quantum gravity as formulated by Bombelli, Lee, Meyer and Sorkin [2] can be seen as a historic culmination of these ideas, and remains the core of the modern CST program.

What do we mean by quantising the causal structure poset, and how can this be enough to recover *all* the information in the spacetime, including the volume element? CST posits that there are a finite number of fundamental spacetime events associated with any finite spacetime volume. This renders the spacetime causal structure poset into a locally finite poset which we call a **causal set**. Thus, spacetime is no longer considered to be continuous at a fundamental level, but rather, atomistic or discrete. In other words, continuum spacetime itself is emergent, arising as an approximation from the discrete substructure, much like the bulk properties of a fluid which is fundamentally molecular in nature.

Taking its cue from the HKMMKP theorem, CST also demands that there must be a correspondence between the number of causal set elements in a spacetime region, to its volume, in units of the discreteness scale. Thus, cardinality or "number" replaces the volume element, while the causal structure is contained in the partial ordering. This gives us the CST slogan

Order + Number $\sim$ Geometry

where $\sim$ represents the continuum approximation. Importantly, this slogan tells us that CST isn't *just* a quantum theory of locally finite posets, but demands also an explicit number to volume correspondence in the continuum approximation. This is critical to its very formulation and distinguishes it from other poset approaches to quantum gravity like those proposed in [19]. Causal sets that have a continuum approximation are said to be **continuumlike**.

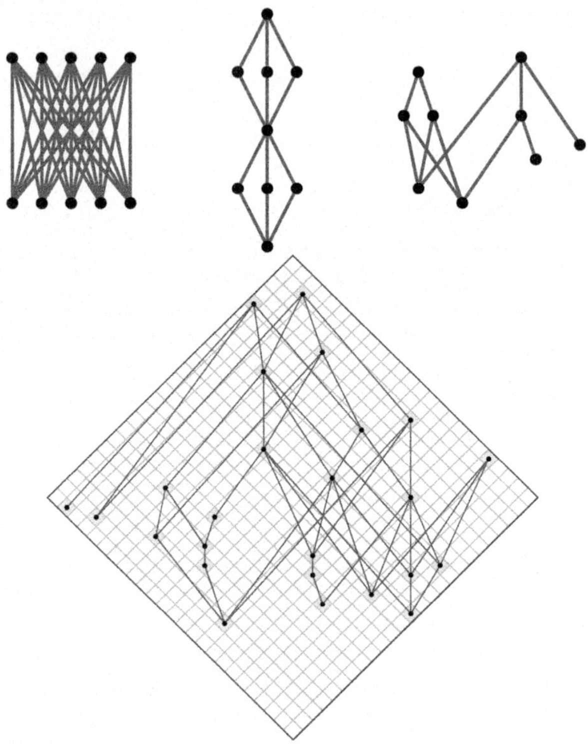

Fig. 1.1 Examples of finite element causal sets represented by their Hasse diagrams, where only the nearest neighbours are depicted. To avoid clutter, the arrows on the relations are not drawn and it is assumed that precedence follows the y-axis of the page. The last figure depicts a 25 element causal set approximated by a causal diamond $\mathbb{D}^2 \subset \mathbb{M}^2$. Appendix A contains many more examples of causal sets

In Fig. 1.1 we see examples of causal sets, including one that is approximated by a region of $d = 2$ Minkowski spacetime $\mathbb{M}^2$.

By assuming a fundamental discreteness, CST proposes that the continuum doesn't exist in all its detail, but only in a coarse grained sense. This combination of causality and spacetime discreteness therefore leads us to a profound reimagining of Lorentzian spacetime, which in turn provides a distinctive, order theoretic foothold on the uphill task of quantising spacetime.

Adopting causality as a principle in quantum gravity, where spacetime itself is dynamical, may seem puzzling and even contradictory. In CST, it is not causality per-se that is assumed, but rather the underlying order theoretic ontology which can be thought of as a "proto-causality" condition. Classical, or semi-classical, spacetime causality is assumed to be emergent from this proto-causal structure as prescribed by the continuum approximation of CST. It is important at this point to note that continuumlike causal sets are rare—indeed, as we will discuss in this book, an overwhelmingly large number of causal sets are nothing like spacetime [20–22]! How causality manifests itself in the deep quantum regime, far away from this approximation is not fully understood, even in CST. Quantum superpositions of causal sets naturally imply that there is no *single* background causal structure and hence that spacetime causality itself is dynamical. Thus, in a quantum theory of causal sets, one needs a more sophisticated notion of "quantum causality". We will see hints of this in the quantum sequential growth models in Chap. 5 (see [23] for recent work on this, as well as [24] for a philosopher's take.)

Whatever the choice of kinematic structure, the elephant in the quantum gravity room is the problem of quantum interpretation. While some might insist that the correct act of quantisation will lead the way, or that we live in multiverses that sprout into uncountably infinite possibilities at uncountably many times, we cannot avoid the fact that quantum theory, as currently formulated, relies for its interpretation on external observers. Quantum gravity, and more specifically quantum cosmology, doesn't allow such a dichotomy between the system and the observer. As Bell eloquently stated in his paper, "Quantum mechanics for cosmologists" [25]:

"Was the world wave function waiting to jump for thousands of millions of years until a single-celled living creature appeared? Or did it have to wait a little longer for some more highly qualified measurer –with a Ph.D.?"

This is not the book to deal with this rich subject. However, towards the end, when we are well equipped with the kinematical tools of CST, we will discuss a decoherence functional based, quantum measure approach [26–32] to describe a quantum sequential growth dynamics for causal sets. Here we can use at least one "quantum truth", the **principle of preclusion**, namely that a set of zero quantum measure does not happen.

The chapters in this book are organised for a pedagogical introduction to the subject. We now give a brief overview of each of these chapters. It will be necessary to resort to terminology that may not be familiar to the reader just as yet, but these will be defined in subsequent chapters. The hope is to provide a summary of the broader structure of the theory, and we suggest using this chapter as a guide as the reader makes their way through the rest of the book.

We present the CST hypothesis in Chap. 2. In Sect. 2.2 we begin with some basic definitions related to the causal structure poset. We then describe some of the fascinating results in Lorentzian geometry which show the intimate relationship between spacetime geometry and the causal structure poset, culminating in the HKMMKP theorem. The CST hypothesis is then presented as a twin statement: the first, on the fundamental discrete ontology of spacetime, and the second, on the nature of the continuum approximation.

The continuum approximation can be thought of as a "first-quantisation", to do with the fundamental kinematic structure that underlies every causal spacetime. It is a prescription for discretising the causal structure poset of a continuum spacetime by making it locally finite, while maintaining the number to volume correspondence as suggested by the HKMMKP theorem. This correspondence cannot be obtained from a regular lattice, since the latter is not covariantly defined. Instead, discretising the spacetime via a random Poisson process at uniform density ρ with respect to the spacetime volume generates a random lattice in which the correspondence holds in the mean. Thus, the spacetime continuum is replaced with an **ensemble** of locally finite posets or causal sets generated via a Poisson sprinkling at uniform density ρ. Although the continuum is used to generate the ensemble, it no longer has relevance once this is done.

The "table-cloth-trick" is a useful visual to keep in mind to understand the continuum approximation. Consider an arrangement of bowls laid out on a table cloth, which is itself placed on a much larger table. If one is skilled enough, the table-cloth can be whisked from under the arrangement, without changing it. Suppose the bowls are randomly arranged at uniform density ρ, and we were to do this trick a multiple number of times, then this would generate an ensemble of random bowl arrangements at that density. With a little imagination, and using the set of relative positions of the bowls, one can try to "reconstruct" where the table-cloth might have been, at scales sufficiently larger than ρ^{-1}. What we cannot capture, are details (folds or holes) in the fabric of the order of, or smaller than ρ^{-1}. The reconstruction thus smoothes away these details which are deemed irrelevant. While this analogy may be confusing to some (we still have the larger table underneath!) it is meant to illustrate the idea behind the continuum approximation of causal sets, which is often baffling to the newcomer.

We perform a similar table-cloth-trick in constructing a continuum approximation in causal set theory. We use the "fabric" of spacetime on which to lay down our random causal sets at a given uniform spacetime density ρ, and then we whisk the spacetime away. All that remains is a causal set, and by repeating the process multiple times, we obtain an ensemble of causal sets which represents the continuum approximation. This is discussed at length in Sect. 2.3.2. We then discuss the distinctive features of causal set discreteness—nonlocality in Sect. 2.3.2.1, voids in Sect. 2.3.2.2 and, importantly, Lorentz invariance in Sect. 2.3.2.3. The **covariant discreteness scale** ρ^{-1} plays an important role in the continuum-approximation. By varying it, one captures different coarse grained features of the spacetime, as discussed in Sect. 2.3.2.4.

Although this ensemble of causal sets is generated from the spacetime, the hope is that every "relevant" feature in the continuum at scales larger than the discreteness scale, can be captured by it. This hope is formalised into the **fundamental conjecture of CST**, its **"Hauptvermutung"**, discussed in Sect. 2.3.3.

Chapter 3 is devoted to continuumlike causal sets, or causal set kinematics. Section 3.2 gives the broad outlines for the process of "geometric reconstruction" from an ensemble of continuumlike causal sets. Just as in the table-cloth-trick, the ensemble of causal sets is generated from the continuum, but the latter is then discarded. The emergent properties of the spacetime are then reconstructed from this underlying order theoretic substructure. This means that we no longer need the continuum to think about causal sets, although it is of course an important approximation to check our reconstruction against. Next, we focus on geometric and topological properties that are reconstructed from three classes of order theoretic structures, namely the chains in Sect. 3.3, antichains in Sects. 3.4, and order intervals in Sect. 3.5. These include estimators for the spacetime dimension [18,33], the spatial homology [34], time-like distances [35] and the discrete Einstein-Hilbert or Benincasa-Dowker-Glaser (BDG) action [36,37], amongst many others. These results prove weaker, observable-dependent versions of the Hauptvermutung, and together, give supporting evidence for it.

An important outcome of the exercise of geometric reconstruction is that it provides a host of observables for CST in the form of **geometric order invariants**, which are a special class of label independent quantities associated with the causal set.

Before we leap into the full quantum dynamics of CST, it is interesting to ask how quantum fields behave on this first quantised spacetime, i.e., on continuumlike causal sets. Chapter 4 is devoted to studying free quantum scalar fields on causal sets. Here, we follow the Sorkin-Johnston procedure for quantising the free scalar field on the causal set via the covariantly defined Sorkin-Johnston (SJ) vacuum [38–40], starting from the discrete retarded and advanced Green's function. In Sect. 4.2 we begin with a broad overview of scalar field Green's functions on causal sets. Explicit constructions of these Green's function in dimension 2 and 4 follow; in Minkowski spacetime in Sect. 4.2.1, in Riemann Normal Neighbourhoods in Sect. 4.2.2, and in de Sitter and anti-de Sitter spacetime in Sect. 4.2.3. In Sect. 4.3 we outline the causal set construction of the SJ vacuum starting from the Green's functions. On a large causal set, the SJ vacuum cannot be calculated analytically, and we must resort to numerical simulations. In $\mathbb{M}^2$ and $\mathbb{M}^4$ as well as $d = 2$ de Sitter spacetime, there is a good agreement with the continuum Minkowski vacuum [38,39]. However, in $d = 4$ de Sitter spacetime, numerical simulations suggest that there are significant modifications to the Bunch Davies vacuum in the deep UV that arise due to discreteness and nonlocality [41]. These are potentially of phenomenological interest since modifications in the quantum field theory vacuum could manifest themselves in the initial conditions for the observable universe.

In Sect. 4.4 we define Sorkin's spacetime entanglement entropy (SSEE) [42] between causally complementary regions in a continuumlike causal set [43,44]. This can be calculated numerically using the SJ vacuum in the cases described above. As expected, with a UV finite theory, the SSEE is finite. However, there is an unexpected surprise: in all these cases, the SSEE on the causal set satisfies a volume rather than an area relation. In [43] it was shown that if an additional truncation is imposed on the discrete spectrum, then the SSEE obeys an area law. One can interpret this result in one of two ways: either the quantum field theory cannot be treated independently of the causal set in the deep UV, or that an area law is not fundamental, but rather, emergent. These possibilities need further, detailed exploration and could provide important insights into the UV behaviour of such quantum field theories. This chapter lays the ground work for the ongoing study of interacting quantum fields on causal sets [45–47] as well as for future work on more general classes of field theories.

Equipped with the kinematic tools we have gathered, in Chap. 5 we confront the main task at hand, which is to construct a fully dynamical quantum theory of causal sets. Partly because of the nonlocal nature of discretisation in the continuum approximation, but also because causal sets are fundamentally Lorentzian in character, it is most natural to quantise causal sets using the path integral or sum-over-histories(SOH) formulation.

Since CST posits that the continuum is only emergent, at a more fundamental level we must include causal sets that are not continuumlike. Thus, in CST we must replace the space of continuum spacetime geometries, or histories, with the space of all causal sets. This is the "second", or dynamical quantisation of causal sets. Using the discrete Einstein-Hilbert or BDG action, the CST partition function $\mathcal{Z}_\Omega$ is defined over a "sample space" Ω of causal sets. Alternatively, the BDG action can be used to define the "double path integral" $\mathcal{D}_\Omega$ using the decoherence functional. As we will see later on, this formulation helps pave the way for a measure theoretic interpretation of quantum theory. One may wish to summarise the problem of causal set quantum gravity as the problem of "evaluating" $\mathcal{Z}_\Omega$ or $\mathcal{D}_\Omega$.

In Sect. 5.2 we start by defining the CST partition function $\mathcal{Z}_{\Omega_n}$ over Ω_n, the set of all n-element causal sets. The cardinality n of a causal set is analogous to the spacetime volume V in the continuum approximation. Hence, as in unimodular gravity [48,49], fixing n is equivalent to fixing the spacetime volume. In Sect. 5.2.1 we discuss the "entropy problem" in CST. Theorems in poset theory show that for large n, Ω_n is overwhelmingly dominated by a class of non-continuumlike causal sets, the Kleitmannn-Rothschild (KR) posets, which have just three "moments of time" [20]. In fact, there is a whole subdominant class of posets with K-moments of time, for $K \ll n$, which are also not continuumlike [21,22,50]. The overwhelming "entropy" of this class of non-continuumlike causal sets poses a challenge to the recovery of continuumlike causal sets for large n, and hence for large spacetime volumes. Such entropic problems are not special to causal sets—a well known example are the branched polymers of dynamical triangulations [51].

It is noteworthy that CST has made significant progress on this problem and the situation may not be so dire. It was shown in a series of papers [52–55] that in the large volume or n limit, the BDG action in any dimension is sufficient to strongly suppress the contribution of such causal sets to $\mathcal{Z}_{\Omega_n}$, when the discreteness scale is larger than or of order the Planck scale! Thus, continuumlike causal sets stand a chance against this entropic dominance.

In general, evaluating $\mathcal{Z}_{\Omega_n}$ is not straightforward except for very small n, and we must turn to numerical methods. This is the focus of Sect. 5.2.2. Typical numerical methods are adapted to either Euclidean and/or statistical partition functions. Since there is no meaningful way to perform a Wick rotation of a causal set to make it Euclidean, a statistical partition function for causal sets must be obtained by alternative means. An analytic continuation of the measure in $\mathcal{Z}_{\Omega_n}$ does the trick. This gives a Lorentzian statistical partition function over causal sets which is amenable to numerical, Markov-Chain-Monte-Carlo (MCMC) simulations, in which expectation values of geometric and other order invariants can be calculated [56–61].

Simulations over the uniform sample space Ω_n (i.e., without the BDG action term) show that the onset of the KR posets occurs around $n \sim 84$ [58]. However, the MCMC moves are in general inefficient because of the inherent nonlocality in causal sets, and this additionally impedes "thermalisation", i.e., the transition to the ground state. One can therefore simplify the problem by restricting the full sample space Ω_n to a subset. In [56,61] topologically and dimensionally restricted subsets of Ω_n were studied in $d = 2$ and $d = 3$ dimensions. Using geometric and other order invariants as order parameters, it was shown using numerical MCMC techniques that all these systems exhibit a first order phase transition as a function of the analytic continuation parameter β [56,59–61]. The transition is from a continuumlike phase at high "temperatures", $\beta^{-1} > \beta_c^{-1}$, to a non-continuumlike layered phase at low "temperatures", $\beta^{-1} < \beta_c^{-1}$. Though it is non-continuumlike, the low temperature phase has the physically interesting feature of being spatially homogeneous due to a high degree of causal connectivity, thus mimicking the initial conditions of our early universe. In the $d = 2$ case the Hartle-Hawking no-boundary wavefunction shows explicit peaks over the two phases [57], suggesting a truly quantum phase transition.

While the numerical simulations are sometimes a necessary and satisfying way to proceed, they nevertheless leave some important open questions in their wake. The most relevant one perhaps, is how to interpret these results after analytically continuing back to the full quantum partition function. The Hartle-Hawking wave function calculation in $d = 2$ gives a partial answer to these questions, but we do not have a general prescription for calculating it in $d = 4$ dimensions.

Important questions also remain when dealing with the full quantum partition function. Apart from technical difficulties in performing exact analytically calculations, we do not know how to meaningfully interpret expectation values of observables in this "single-path" language, without the use of external observers. Moreover, we have used a continuum inspired action, and one might legitimately ask whether there is a more fundamental, order theoretic measure in the path integral.

In Sect. 5.3 we present the **sequential growth** paradigm of Rideout and Sorkin [62] which is a more order theoretic, foundational approach to dynamics based on purely order theoretic principles. It is a radical departure from the continuum-inspired dynamics and hence presents unique challenges. In Sect. 5.3.1 we present the basics of this growth dynamics and focus on the future growth models, which generate a sample space of past finite countable causal sets [62–64]. We begin in Sect. 5.3.1.1 with a refresher of the measure theoretic structures underlying the $1 + 1$ dimensional classical random walk. In Sect. 5.3.1.2 we use this as a guide to set up the **event algebra** for the sequential growth models [65]. In this framework the covariant observables (or beables) are the elements of the covariant event algebra. The measure, whether classical or quantum, is then defined over this algebra.

In Sect. 5.3.2 we present the well-studied Rideout-Sorkin classical sequential growth models [62,63,66]. These models provide a template for the quantum sequential growth introduced in Sect. 5.3.3 where the classical measure is replaced with a **quantum measure** [27,28,67–71]. Using the double-path integral and its associated **decoherence functional**, the quantum measure takes values in a **histories Hilbert** space constructed from the event algebra [69]. Not every quantum sequential growth model is covariant—one must do more work to prove that such models exist. This can be demonstrated in a simple class of models [71]. It is a hard, but important task to extend these results to more realistic dynamical models, and additionally, to calculate the quantum measure for physically relevant observables. Exploring and understanding all aspects of the quantum sequential growth paradigm and the emergence of the classical regime is what a full theory of causal sets quantum gravity aspires to.

A few words about the quantum measure. It can be viewed as a first step towards an observer independent quantum theory. Without going into deeper interpretational questions, the quantum measure already allows, at the minimum, the following simple interpretation via the **principle of preclusion**, namely that sets of zero quantum measure do not occur. Hence, in this possibly limited sense, the interpretation is predictive, without recourse to external observers.

We wind up in Chap. 6 with a brief discussion on CST phenomenology in Sect. 6.2. Perhaps the best *physical* justification for studying CST is that it has contributed to two important predictions for quantum gravity phenomenology. We have already mentioned the first, namely that despite discreteness, CST does not violate local Lorentz invariance in the continuum approximation [72]. This is consistent with observations which have tested Lorentz invariance to great accuracy [73]. The second, which we present in Sect. 6.2 is Sorkin's *accurate* 1987 prediction for the value of the cosmological constant Λ [48,74–76]. Strictly, the prediction is for the *fluctuation* $\Delta\Lambda$ about an assumed value of $\Lambda = 0$. Models of fluctuating Λ have been studied extensively and are particularly interesting in light of recent reports of a "Hubble tension" [76–80]. In these models the dark energy density is **ever present** in that it tracks the critical density at all times. While the preservation of local Lorentz invariance is a mathematically rigorous result, the more detailed predictions for $\Delta\Lambda$ come from complicated phenomenological model building which is beyond the scope of this book. Instead we point the interested reader to the literature for the most recent developments [79,80]. The brevity in the treatment of phenomenology in this book should in no way reflect its importance in CST, but rather this author's limitations. CST phenomenology is a key aspect of the theory, and covering this vast topic is a book in its own right!

We end this chapter, and the book, in Sect. 6.3 with a list of open questions and directions in CST. The list is necessarily incomplete, but spring from aspects of CST we have already discussed in the book.

Over the past couple of decades of my engagement with CST, I have encountered questions of the following type. "What is the evidence for spacetime discreteness?" I would counter this by asking, "what is the evidence for the continuum?" None of our spatio-temporal experiments or observations (even numerically) are done truly continuously. Continuum calculations make fundamentally discrete calculations easier to do, with powerful tools like calculus and analysis, but this calculational ability has led to a philosophical mindset that the continuum is an absolute certainty. In CST, the fundamental discreteness scale $\ell = \rho^{-1/d}$ must be determined by the internal consistency of the theory as well as with observation. An example of the former is the suppression of entropically dominant, non-continuumlike contributions by the BDG action, discussed in Chap. 5, while an example of the latter is Lorentz invariance, the value of $\Delta\Lambda$, and the non-existence of voids.

Exercises are interspersed throughout the book. These vary in simplicity—some are quite basic and simply "fun", while others deal with slightly more subtle aspects of the theory. Like graph theory, causal sets can be quite addictive to work with. Its always a pleasant surprise to find that relatively small causal sets contain rich continuum geometric information. In addition to analytic calculations, we encourage those interested in numerical methods to try out simulations of causal sets. As we wind our way through the chapters of the book, we will end up at several "frontiers" of CST.[1] These are roughly the boundaries of our current knowledge, and in some cases, areas of current exploration. I hope that the journey, and these questions, will entice readers into actively thinking about and further, working in CST.

I have often been asked by undergraduate and graduate students if there is an introductory book on CST. This is the closest I have been able to manage, and I hope students will find it a useful hook into an exploration of the theory. I have tried to retain a certain level of pedagogy for this reason. Basic prerequisites are the same for any book on quantum gravity. It is a definite advantage to have a solid understanding of general relativity, especially some of its underlying mathematics, as well as of quantum field theory.

A disclaimer: this book does *not* cover every aspect of CST. The aim is more to give a broad outline of the theory with enough detail in some cases for the sake of pedagogy. Since I am an active participant in this research, there is, undoubtedly, also bias in the presentation and the choice of examples. The reader should therefore not treat this book, in any sense, as the last word on the subject, but rather as a first introduction to CST. It is also not meant for the seasoned causal set worker, except possibly to provide easy access to the CST bibliography.

[1] The term "we", which I will use now and then in this book, indicates what I hope is a shared journey with the reader.

2 The Causal Set Hypothesis

Abstract

This chapter is a technical introduction to CST. It begins with a review of the causal structure for causal spacetimes, which is a poset, and leads up to a set of theorems in Lorentzian geometry which we collectively refer to as the HKMMKP theorem. A discrete version of this theorem provides the primary motivation for the CST hypothesis and helps shape its continuum approximation, which is conjectured to be unique. This approximation exhibits several interesting features, including Lorentz invariance (despite discreteness), and a certain characteristic nonlocality. This chapter also contains definitions and notations which will be used in the remainder of the book.

2.1 Introduction

In this chapter the basic tenets of CST are laid out in detail. As stated in Chap. 1, this approach is rooted in Lorentzian geometry and built on the principles of causality and discreteness. The fundamental ontology of CST is a **causal set**, which is a locally finite poset. The order relation in the causal set represents a "proto-causality" relation, which is realised as a spacetime causality relation only in the continuum approximation of the theory.

At the most fundamental level CST is a quantum theory of causal sets, an overwhelming number of which are nothing like spacetime [20,21,50]. The set of theorems we collective refer to as the HKMMKP theorem tethers CST to Lorentzian geometry by allowing us to define, in a covariant manner, when a causal set is approximated by a continuum spacetime or is **continuumlike**.

In this chapter we present the CST hypothesis and define its continuum approximation. In subsequent chapters we will flesh out CST in detail: (i) the extraction of observables or geometric and other order invariants in causal sets which have a continuum approximation (Chap. 3), (ii) the construction of quantum fields on continuumlike causal sets (Chap. 4) (iii) the construction of a full quantum dynamics for

S. Surya, *The Causal Set Approach to Quantum Gravity*, Lecture Notes in Physics 1036,
https://doi.org/10.1007/978-3-031-84420-1_2

causal sets including those that are not continuumlike, and finally (iv) a discussion on key phenomenological results in CST.

Because of the deep connection between CST and Lorentzian geometry, we start with a discussion in Sect. 2.2 on the importance of the causal structure in Lorentzian geometry. Section 2.2.1 provides some definitions and some background for the study of causal structure. Section 2.2.2 discusses some of the key results on the role of causal structure in Lorentzian spacetime—from Robb's definition of the "conical order", to the Alexandrov-Ovchinnikova-Zeeman results on Minkowski spacetime, $\mathbb{M}^d$, to the Kronheimer-Penrose axiomatisation of causal structure and finally, to the HKMMKP theorem.

This brings us to the Bombelli-Lee-Meyer-Sorkin proposal for CST [2] in Sect. 2.3. By discretising the causal structure poset to produce a locally finite poset, the continuum approximation of CST is defined explicitly using a discrete realisation of the HKMMKP theorem. While there have been similar ideas on quantising the causal structure [1, 15, 81], it is the CERN preprint of Myrheim's [18] that comes closest to the CST proposal.

In Sect. 2.3.1 we set down much of the notation that will be used in the rest of the book. Section 2.3.2 is devoted to the continuum approximation, which is essential to CST and distinguishes it from other poset-based approaches. The CST "random lattice" differs significantly from a regular discretisation of spacetime. Using the example of a flat $d = 2$ causal diamond or Alexandrov interval $\mathbb{D}^2 \subset \mathbb{M}^2$, we discuss various properties of the continuum approximation. The combination of discreteness and covariance gives rise to a characteristic nonlocality described in Sect. 2.3.2.1, which leaves its imprint in virtually every aspect of the theory. The randomness also means that there is a non-zero probability of "voids" where the continuum breaks down completely. We point to a calculation of the probability of nuclear size voids and what its implication is for the large time state of the universe in Sect. 2.3.2.2. In Sect. 2.3.2.3 we present the very important result that, despite discreteness, there is no violation of Lorentz invariance [72]. Section 2.3.2.4 discusses the role of coarse graining in CST which relates different scales in the theory, in a manner similar to blocking transformations in the Ising model.

In Sect. 2.3.3 we state the fundamental conjecture or "Hauptvermutung" of CST, which gives a kinematic justification for replacing continuum spacetime with causal sets. We will see ample evidence in Chap. 3 to support this conjecture.

We end with a brief summary of the chapter in Sect. 2.4.

2.2 Spacetime Causal Structure

2.2.1 Basics

The reader who is familiar with the basics of causal structure from the book of Hawking and Ellis [82] or Penrose's monographs [83] may skip this subsection.

A **Lorentzian spacetime** is the pair (M, g), where M is a connected, paracompact, time orientable, C^3, d-dimensional manifold, and g is a Lorentzian metric, i.e.,

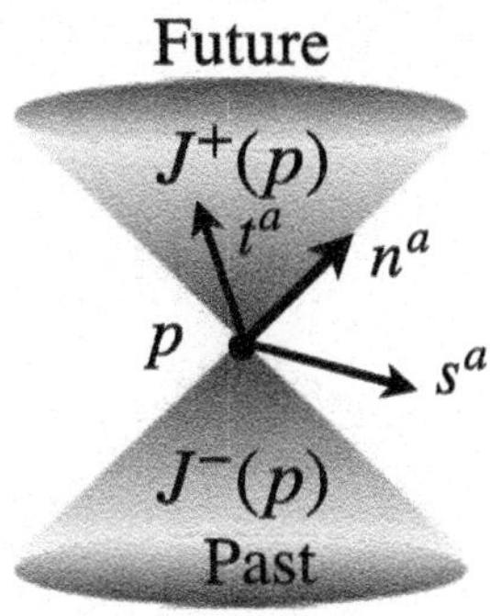

Fig. 2.1 The light cone in the tangent space at an event $p \in M$ divides the set of tangent vectors into those that are spacelike s^a, timelike t^a and null n^a. The latter two vectors come with an additional label of being either future directed or past directed, depending on whether they belong to the future or the past light cone of p

of signature $(-, +, +, \ldots,)$. We denote d-dimensional Minkowski spacetime by $\mathbb{M}^d \equiv (\mathbb{R}^d, \eta)$, where $\eta_{ab} dx^a dx^b = -dx_0^2 + \sum_{i=1}^{d-1} dx_i^2$.

At every event $p \in M$ the tangent space $V_p \simeq \mathbb{M}^d$ and hence splits up into **timelike** ($\mathrm{v}^a \mathrm{v}_a < 0$), **spacelike** ($\mathrm{v}^a \mathrm{v}_a > 0$) and **null** vectors ($\mathrm{v}^a \mathrm{v}_a = 0$.) A vector $\mathrm{v}^a \in V_p$ is said to be **causal** if it is either timelike or null and is **future or past directed** with respect to the chosen time-orientation on M. Given a time-orientation on (M, g), the set of all causal vectors split into a **future** and **past lightcone** in V_p which intersect only at the origin (see Fig. 2.1.) This topological "double cone" structure is essential for a definition of causality, since a past directed causal vector cannot be continuously deformed to a future directed vector while remaining causal, and vice versa.

This tangent space structure can be used to define causal relations between events in the spacetime. The identification of causal vectors in V_p allows us to characterise special classes of C^2 curves on (M, g). A C^2 curve γ is said to be **future timelike (FTL)** if its tangent at every point $p \in \gamma$ is timelike future directed and similarly, it is said to be **past timelike (PTL)** if its tangent is everywhere timelike past directed. **Future causal (FC) and past causal (PC)** curves are similarly defined. Associated with every event $p \in M$ therefore, is its **chronological future set** $I^+(p)$**, and past set** $I^-(p)$ and its **causal future set** $J^+(p)$ **and past set** $J^-(p)$**:**

$$\begin{aligned}
I^+(p) &\equiv \{q \in M \mid \exists\ FTL\ \gamma : [0,1] \to M, \gamma(0) = p, \gamma(1) = q\} \\
I^-(p) &\equiv \{q \in M \mid \exists\ PTL\ \gamma : [0,1] \to M, \gamma(0) = p, \gamma(1) = q\} \\
J^+(p) &\equiv \{q \mid \exists\ FC\ \gamma : [0,1] \to M, \gamma(0) = p, \gamma(1) = q\} \\
J^-(p) &\equiv \{q \in M \mid \exists\ PC\ \gamma : [0,1] \to M, \gamma(0) = p, \gamma(1) = q\}. \qquad (2.1)
\end{aligned}$$

The sets $I^\pm(p)$ are open sets in the manifold topology, and while $J^\pm$ need not be closed, $\overline{I^\pm(p)} = \overline{J^\pm(p)}$. An **Alexandrov interval** is the "causal diamond" (Fig. 2.2)

$$I(p, q) \equiv I^+(p) \cap I^-(q), \qquad (2.2)$$

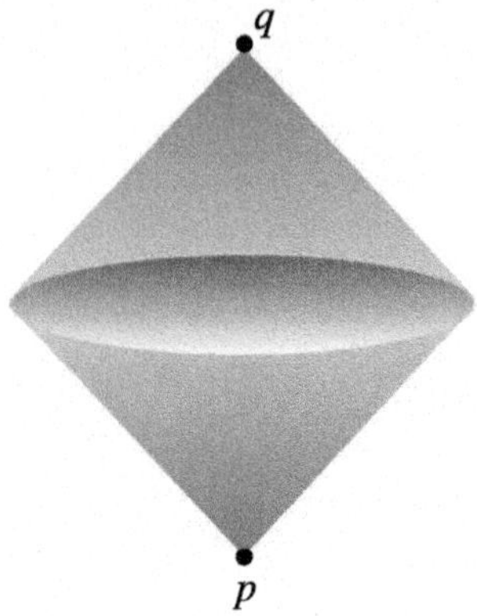

Fig. 2.2 A causal diamond or Alexandrov interval $I(p,q) = I^+(p) \cap I^-(q)$. This is an open set in the manifold topology when the spacetime is C^3 differentiable

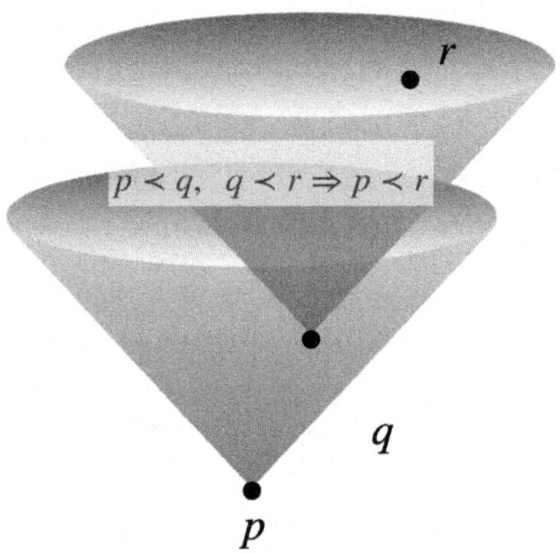

Fig. 2.3 The nesting of lightcones shows the transitivity of the relations $\prec$ and $\prec\!\!\prec$

and is therefore an open set in the manifold topology. We will use the notation $\mathrm{vol}(p,q) \equiv \mathrm{vol}(I(p,q))$ to denote the spacetime volume of $I(p,q)$ with respect to the metric g_{ab}.

From the definitions above, $q \in I^+(p) \Leftrightarrow p \in I^-(q)$ and similarly, $q \in J^+(p) \Leftrightarrow p \in J^-(q)$, which suggests an ordering relation between events in (M, g). The **causal relation** $\preceq$ and the **chronological relation** $\prec\!\!\prec$ relation in (M, g) are defined as

$$
\begin{aligned}
p \preceq q &\Leftrightarrow q \in J^+(p),\, p \in J^-(q) \\
p \prec\!\!\prec q &\Leftrightarrow q \in I^+(p),\, p \in I^-(q)
\end{aligned}
\tag{2.3}
$$

These relations are **transitive**: $p \prec q, q \prec r \Rightarrow p \prec r$, and $p \prec\!\!\prec q, q \prec\!\!\prec r \Rightarrow p \prec\!\!\prec r$ (see Fig. 2.3) since the union of the causal or timelike future directed C^2 curve from p to q and from q to r can be deformed into being C^2 also *at* q.

Exercise 2.1 Consider the $d \geq 3$ dimensional geometries

$$g_{ab}dx^a dx^b = -\sum_{i=1}^{m} dt_i^2 + \sum_{j=i}^{d-m} dx_j^2. \tag{2.4}$$

Find the surface that the null vectors sweep out from the origin. Use this to show that only when $m = 1$ are there natural order relations $\prec$ and $\prec\!\!\prec$ on the set of events. What is the lightcone structure for $m > 1$, and what might replace the sets $I^{\pm}$, $J^{\pm}$?

Given $\prec$ and $\prec\!\!\prec$ on (M, g), the **horismotic** relation $\rightarrow$ is

$$p \rightarrow q \Leftrightarrow p \prec q,\ p \not\prec\!\!\prec q \tag{2.5}$$

which is the "null" or lightlike relation. This relation is *not* transitive, as can be checked in the simple case of $\mathbb{M}^2$. To complete the set of all relations we can also define the **spacelike** "un-relation" $\|$ as

$$p \parallel q \Rightarrow p \npreceq q \text{ and } q \npreceq p. \tag{2.6}$$

As the reader can verify, this relation is not an order relation (except trivially, in $\mathbb{M}^2$.)

Even though causal and chronological curves are defined via the light cones in the tangent space at every point, it is important to emphasise that the relations $\prec$ and $\prec\!\!\prec$ are global in character. Because of this, it is useful to distinguish between local and global causality. For any open neighbourhood U of p let $I^{\pm}(p, U)$ and $J^{\pm}(p, U)$ be defined as in Eq. (2.1), but for chronological and causal curves that lie wholly inside the "subspacetime" $(U, g) \subset (M, g)$, with the causal and chronological relations $\prec_U$ and $\prec\!\!\prec_U$, respectively. These local chronological and causal relations do not suffice to determine the corresponding global relations, even for events that lie within U. In general, $I^{\pm}(p, U) \subseteq I^{\pm}(p) \cap U$ and $J^{\pm}(p, U) \subseteq J^{\pm}(p) \cap U$ since causal curves in (M, g) may leave and re-enter U. Importantly, global violations of causality are possible, even when there is no local violation. The simplest example of such a spacetime is $\mathbb{M}^d$ with a time-identification $t \sim (t + T)$, $T > 0$, while a more sophisticated example is the Gödel spacetime, which possesses causality violating regions [82,84]. At such events, causal curves leave and reenter local neighbourhoods giving rise to **closed causal curves**, favoured by time-travel enthusiasts, but which we disfavour in CST on physical grounds.

Since causality is taken to be a fundamental principle in CST, we restrict our attention to spacetimes for which $\prec$ and $\prec\!\!\prec$ are **acyclic**:

$$\begin{aligned} p \preceq q,\ q \preceq p &\Rightarrow p = q \\ p \prec\!\!\prec q &\Rightarrow q \not\prec\!\!\prec p. \end{aligned} \tag{2.7}$$

Such spacetimes are referred to as **causal spacetimes**. The assumption of causality leads to a mathematical object which is at the heart of this book, namely, a partially ordered set or **poset**. This is a set with a (directed) order relation $\prec$ on its elements which is both acyclic and transitive. Thus, underlying every causal spacetime (M, g) is a **causal structure poset** $(M, \prec)$, and the **chronological structure poset** $(M, \prec\!\!\prec)$. Forgetting where these posets come from, we can treat M simply as a *set* of events (i.e., sans differential or topological structure) with a partially ordered relation ($\prec$ or $\prec\!\!\prec$.) In CST $(M, \prec)$ is the primary object which we wish to "quantise" and hence it is important to understanding its role in Lorentzian geometry as best as possible.

Henceforth we will assume that spacetime is causal. However, even if causality is strictly satisfied, there are other global causal properties of (M, g) which, if violated, can render a spacetime physically pathological. This gives rise to a hierarchy of global causality conditions, one stricter than the other [82, 83, 85].

In order to define these properties, we need to consider a special class of neighbourhoods. Locally, the causal structure of every Lorentzian spacetime looks like an open ball in $\mathbb{M}^d$. More precisely, there exists a bijective map $\exp_p$ from an open neighbourhood O_p of the origin in V_p to a neighbourhood U_p of p in M such that: each vector $v^a \in V_p$ is the tangent of a geodesic from p, and traversing it at unit affine parameter (given by the length of v^a) defines a unique event $q \in U_p$. Clearly, O_p and U_p can be as small as possible, but in a generic spacetime, the uniqueness of the map breaks down when geodesics intersect. In particular, $\exp_p$ allows us to think of the local causal structure in U_p as a "small" deformation of that in $\mathbb{M}^d$. Further, one can define a special class of neighbourhoods U_p, the **convex normal neighbourhoods** (CNNs) such that for every $r, s \in U_p$, $I(r, s)_{U_p} \equiv I^+(r, U_p) \cap I^-(s, U_p) \subset U_p$ (see [82, 83] for more details.)

One of the most stringent causality conditions is that of **global hyperbolicity**, which is essential for the well-posedness of the initial value problem [86, 87]. It is the requirement that $\forall\, p, q \in M, \overline{I(p, q)}$ is compact in the manifold topology. There are other, equivalent definitions, but this is the most relevant one for CST.

A far less stringent condition is that of **strong causality**, which prevents a causal curve from coming "too close" to itself after a finite, non-zero change in the parameter along the curve. This protects the spacetime from causality violations by arbitrarily small perturbations of the spacetime. A spacetime (M, g) is said to be **strongly causal** if at every $p \in M$, there exist arbitrarily small CNNs $U_p \ni p$ such that no causal curve that leaves U_p can re-enter it, i.e., $I^\pm(q, U_p) = I^\pm(q) \cap U_p$, $\forall q \in U_p$ and $I(r, s)_{U_p} = I(r, s) \subset U_p$. As a result, in strongly causal spacetimes the **Alexandrov topology** generated from Alexandrov intervals $I(p, q)$ (which are open in the manifold topology) can be shown to be the same as the manifold topology [83].

An even less stringent causality condition is that of **future and past distinguishability**. In such a spacetime (M, g), every distinct pair of events $p, q \in M$, satisfies the conditions $I^+(p) \neq I^+(q)$ and $I^-(p) \neq I^-(q)$. Thus the chronological past and future of any event in (M, g) distinguishes it from all other events. Strongly causal spacetimes are always future and past distinguishing, but the converse is not always true, as in the example shown in Fig. 2.4.

Every $p \in M$ in a future and past distinguishable spacetime possesses arbitrarily small **locality neighbourhoods** $U_p \in p$ in which no future directed causal curves that leave U_p *from* p can re-enter U_p. Such spacetimes are therefore also robust

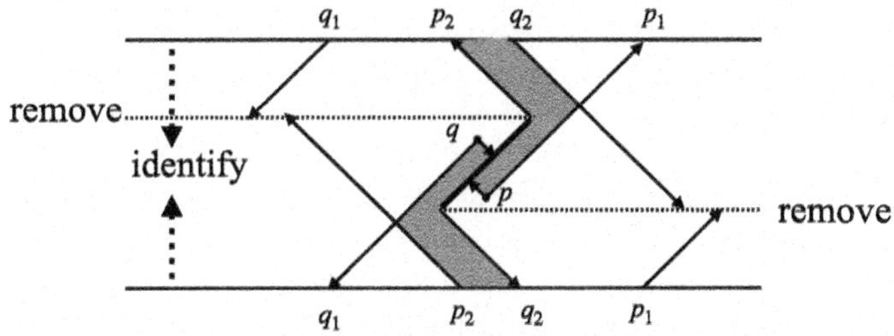

Fig. 2.4 An example of a spacetime which is future and past distinguishing but not strongly causal. The bottom horizontal line is identified with the top horizontal line, but there is no violation of causality or even of future and past distinguishability. Strong causality is violated only along the central null geodesic, shown here as a thickened line. The closure of the Alexandrov interval $I(p, q)$ for p and q that straddle this geodesic is non-compact and non-local

against causality violations from arbitrarily small perturbations, though of course more weakly so than strongly causal spacetimes. Relaxing the causality condition even further would allow either past directed or future directed curves from p to come arbitrarily close to p. Hence, future and past distinguishing spacetimes satisfy the weakest causality condition in the continuum compatible with the physical criterion of causality.

2.2.2 Causal Structure and Lorentzian Geometry

The role of the causal structure in $\mathbb{M}^4$ was identified very early on by Robb in his 1914 book titled "A Theory of Time and Space" and its 1936 edition retitled "Geometry Of Time And Space" [8,88]. He noticed that every event in $\mathbb{M}^4$ has a causal future and causal past, and recognised that this leads to an ordering of events via "before" and "after" relations. He referred to this as a "conical" order on what he called the "elements" of spacetime.[1] This order satisfies a set of postulates, the first and third of which are the very definition of a poset:

- Robb's Postulate I. If an element B be after an element A, then the element A is not after the element B.
- Robb's Postulate III. If an element B be after an element A, and if an element C be after the element B, the element C is after the element A.

The reader will recognise that Postulate I is the acyclic condition, given $A \neq B$, while Postulate 2 is the condition of transitivity. These relations, he noted, are "temporal" and "absorb" the theory of space into a theory of "time".[2] If we substitute "temporal" and "time" for causal, then this may have been the first glimpse into the centrality

[1] We will later use this terminology for causal sets, but it was used by Robb to distinguish spacetime events from actual physical events.

[2] This idea of the primacy of time is also reflected in the work of Synge [89].

of causality in spacetime geometry. As we will see in Chap. 3, spatial structures—both geometric and topological—can be obtained using the causal structure and the volume element.

There seems to have been little attention paid to this insight of Robb's in the ensuing decades, until the early 1950s when Alexandrov and Ovchinnikova [9, 10] first showed a remarkable connection between the causal structure of $\mathbb{M}^d$ and the inhomogeneous Lorentz transformations with dilations. This result appeared in Russian and is largely unknown, but was brought to light thanks to Lester's later work on this subject [90]. A decade later, Zeeman [11] independently came to the same conclusion. Because the earlier Russian articles are inaccessible to this author, we will sketch Zeeman's version of this important result.

We will first need a few more definitions.

A **chronological automorphism** f on $\mathbb{M}^d$ is a bijection $f : \mathbb{M}^d \to \mathbb{M}^d$, such that

$$p \prec\!\!\prec q \Leftrightarrow f(p) \prec\!\!\prec f(q), \quad p, q \in \mathbb{M}^d. \tag{2.8}$$

Similarly, a **causal automorphism** on $\mathbb{M}^d$ is a bijection $f : \mathbb{M}^d \to \mathbb{M}^d$ such that

$$p \prec q \Leftrightarrow f(p) \prec f(q), \quad p, q \in \mathbb{M}^d. \tag{2.9}$$

Theorem 2.1 Alexandrov-Ovchinnikova-Zeeman: *The group $\mathcal{G}_A$ of chronological automorphisms of $\mathbb{M}^d$ is isomorphic to the group $\mathcal{G}_{ILor}$ of inhomogeneous Lorentz transformations and dilations on $\mathbb{M}^d$, when $d > 2$.*

Sketch of Proof: In $\mathbb{M}^d$, $p \to q$, iff (i) $p \not\prec\!\!\prec q$ and (ii) if for every r with $q \prec\!\!\prec r$, it is also the case that $p \prec\!\!\prec r$. For any chronological automorphism f, $p \not\prec\!\!\prec q \Leftrightarrow f(p) \not\prec\!\!\prec f(q)$. This means that for every $f(r) \in I^+(f(q))$, $f(r) \in I^+(f(p))$ (since f is a bijection), and hence $f(p) \to f(q)$. This is also true for f^{-1} and hence f is also a causal automorphism, since it preserves the horismotic relation. The causal automorphisms generate the **causality group** $\mathcal{G}_A$.

The generators of the group $\mathcal{G}_{ILor}$ of inhomogeneous Lorentz tranformations comprise the rotations in the spatial directions, boosts in the time-space directions and translations. Under any of these transformations, $ds^2 \to ds^2$ and therefore the causal structure is preserved. This is also true for dilations $ds^2 \to \lambda^2 ds^2$ since this does not change the loci of the lightcones, and hence $\mathcal{G}_{Lor} \subset \mathcal{G}_A$.

To prove the converse is not as straightforward, and we summarise the main elements of Zeeman's proof. As we have seen, any $f \in \mathcal{G}_A$ preserves the horismotic relation and hence it maps null-rays to null-rays. If $\mathbf{u}_1, \mathbf{u}_2$ are a pair of parallel null-rays in $\mathbb{M}^d$, then it follows that the null rays $f(\mathbf{u}_1)$ and $f(\mathbf{u}_2)$ are non-intersecting. To show that they are also parallel requires some work. We will only sketch a part of this argument below, to give the interested reader a flavour.

Let P denote the plane containing the pair of parallel null rays $\mathbf{u}, \mathbf{u}'$. (a) If P is timelike (its normal is spacelike), then we can use the lightcone coordinates on P to set up a grid of two intersecting parallel families U and V of null rays. Each $\mathbf{u} \in U$ intersects every $\mathbf{v} \in V$, and therefore each $f(\mathbf{u}) \in f(U)$ intersects every

$f(\mathbf{v}) \in f(V)$. The task is to show that $f(\mathbf{u}_1)$ and $f(\mathbf{u}_2)$ are also co-planar, i.e., they are not skew (for $d > 2$.) Assume otherwise, i.e., they span a 3 dimensional (complete) subspace S of $\mathbb{M}^d$ which is spanned by the set $f(U)$ of null rays. Each $f(\mathbf{v}) \in f(V)$ intersects both $f(\mathbf{u}_1)$ and $f(\mathbf{u}_2)$ and therefore also lies in S. This in turn means that every $f(\mathbf{u}) \in f(U)$ lies in S. The two families $f(U)$ and $f(V)$ therefore generate what are known as "doubly-ruled quadrics". These are are surfaces made up of two families of lines where each line in one family crosses all the lines in the other family, and those in the same family are all mutually skew. There are only two types of such quadrics : the hyperboloid and the hyperbolic paraboloid. In an argument (too complex to discuss in detail) Zeeman showed that in both cases this leads to a contradiction as one goes to the asymptotic region. This proves that $f(\mathbf{u}_1)$ and $f(\mathbf{u}_2)$ must be co-planar and non-intersecting and therefore also parallel.

(b) If P is null (the only other option), the argument is much simpler, since there is only one family U of parallel null rays, which are mutually spacelike to each other. Namely, for any $\mathbf{u}_1, \mathbf{u}_2 \in U$, every $p \in \mathbf{u}_1$ is spacelike to every $q \in \mathbf{u}_2$, and similarly, every $f(p) \in f(\mathbf{u}_1)$ is spacelike to every $f(q) \in f(\mathbf{u}_2)$. Assume that $f(\mathbf{u}_1)$ and $f(\mathbf{u}_2)$ are skew. This means that there exists a shortest spacelike distance between $f(\mathbf{u}_1)$ and $f(\mathbf{u}_2)$. Let $p_1 \in f(\mathbf{u}_1)$ and $p_2 \in f(\mathbf{u}_2)$ be the (unique) points which realise this shortest distance. Since these are spacelike to each other, their future and past light cones will eventually intersect in $\mathbb{M}^d$. It then follows that for any $q_1 \in \mathbf{u}_1 \backslash p_1$, with $q_1 \prec p_1$, $\exists\, q_2 \in \mathbf{u}_2, q_2 \succ p_2$ such that $q_1 \prec\!\!\prec q_2$. This is a contradiction.[3]

Finally, f can be shown to be linear, which implies that $\mathcal{G}_A \subseteq \mathcal{G}_{ILor} \Rightarrow \mathcal{G}_A \simeq \mathcal{G}_{ILor}$. Thus, the physical invariants associated with the Lorentz group and dilatations in $\mathbb{M}^d$ follow naturally from its causal structure poset $(\mathbb{R}^d, \prec)$. □

This result suggests that perhaps there is more to the causal structure poset $(M, \prec)$ than meets the eye. As we will see below, this is indeed the case. Under fairly weak causality conditions, $(M, \prec)$ contains the full conformal geometry of (M, g).

In a significant step away from continuum spacetime geometry, Kronheimer and Penrose (KP) [1] defined an axiomatic **causal space** $(X, \prec, \prec\!\!\prec, \rightarrow)$ as one where $(X, \prec)$ and $(X, \prec\!\!\prec)$ are posets and, additionally, for $x, y, z \in X$, (i) $x \prec\!\!\prec y \Rightarrow x \prec y$, (ii) $x \prec y \;\&\; y \prec\!\!\prec z \Rightarrow x \prec\!\!\prec z$ and $x \prec\!\!\prec y \;\&\; y \prec z \Rightarrow x \prec\!\!\prec z$ (the push-forward condition), and (iii) $x \rightarrow y$ iff $x \prec y \;\&\; x \not\prec\!\!\prec y$. The causal structure $(M, \prec)$ of a causal spacetime (M, g) is therefore an example of a causal space. The idea of this axiomatisation is to derive all other aspects of a spacetime from a causal space, which would indicate that the latter is as primitive as the continuum spacetime. The quote at the start of Chap. 1 reflects this spirit, namely that causal spaces can be considered the foundation of spacetime and should therefore be the basis for quantisation. As we will soon see, this is similar in spirit to CST but differs from it since the latter uses only order one relation, $\prec$.

The KP axioms imply that if $x \prec y \prec z$ and $x \rightarrow z$ then $x \rightarrow y \rightarrow z$. The converse conclusion cannot be made. Thus, the horismotic relation, being a boundary

[3] This is a small modification of the proof in [11].

relation, has a certain "rigidity", and captures the fact that that there is a unique (unbroken) null geodesic from x to y if $x \rightarrow y$. The KP axiomatic construction leads to rich secondary structure when some additional requirements are placed on the causal space [1]. This bare-bones axiomatic approach thus holds the key to a causal ontology of spacetime.

In the sketch of the proof of Theorem 2.1 we noted that the causal structure of $\mathbb{M}^d$ is invariant under dilatations. This is trivially true more generally, namely for spacetimes related by C^2 conformal transformations $\widetilde{g}_{ab} = \Omega^2 g_{ab}$, since events that are null related remain so under the transformation. Thus, $(M, \prec)$ remains invariant under conformal transformations. Conversely, if there is a transformation which *locally* preserves the geometric and topology structure of the light cones at every $p \in M$, then it must also preserve the metric *at p* upto an overall scale. Thus, the question we wish to ask is whether causality or chronology preserving transformations also preserve local lightcones.

We define a **causal bijection** between two causal spacetimes as a bijection $f : (M, g) \rightarrow (M, g')$ such that

$$x \prec y \Leftrightarrow f(x) \prec f(y), \tag{2.10}$$

which therefore preserves the causal structure, i.e., $(M, \prec_g) \simeq (M', \prec_{g'})$. Similarly, a **chronological bijection** is a bijection $f : (M, g) \rightarrow (M, g')$ such that

$$x \prec\!\!\prec y \Leftrightarrow f(x) \prec\!\!\prec f(y), \tag{2.11}$$

and preserves the chronological structure, i.e., $(M, \prec\!\!\prec_g) = (M', \prec\!\!\prec_{g'})$. In order to preserve the topological integrity of the local lightcones, a causal or chronological bijection f must additionally be a C^3 diffeomorphism.

Consider a spacetime (M, g) which satisfies the global causality condition of **strong causality** defined in Sect. 2.2.1. For such spacetimes, we noted that the Alexandrov topology is the same as the manifold topology. Under a chronological bijection therefore, f preserves the Alexandrov topology and hence the manifold topology, which means that f is a homeomorphism. For such a spacetime, Hawking [12] proved that

Lemma 2.1 Hawking's Theorem: *If M and M' are n-dimensional C^r manifolds ($n \geq 3, r \geq 3$) with C^{r-1} Lorentz metrics g and g', such that the strong causality assumption holds on M, and if f is a bijection $f : M \rightarrow M'$ such that f and f^{-1} preserve chronological relationships (i.e., a chronological bijection), then f is a C^r diffeomorphism.*

An intermediate step in Hawking's proof is that for strongly causal spacetimes, f is not only a homeomorphism (as pointed out above) but also takes future directed null geodesics in (M, g) to future directed null geodesics in (M', g') (and similarly f^{-1}.) f can then shown to be a C^r diffeomorphism which means that locally, null cones are preserved. This is all we need to show that f is a C^r conformal isometry.

This is the content of Theorem 6 of Hawking, King and McCarthy [13] which completes the statement for strongly causal spacetimes:

Theorem 2.2 Hawking-King-MacCarthy: *A chronological bijection between two strongly causal n-dimensional C^r, $n \geq 3, r \geq 3$ spacetimes is a C^r conformal isometry.*

This means that the differential structure of a strongly causal spacetime M can be obtained from the chronological structure by constructing charts as follows. For any $U \subset M$ there exists a g' on an open subset O of $\mathbb{R}^n$ such that there is a bijection $f' : (U, g) \rightarrow (O, g')$ which is also a causal bijection. This defines a coordinate chart $(U, f'(U))$ on (U, g). Since this can be done for any locally finite cover of (M, g), this gives the requisite atlas on (M, g).

Recognising that the key elements in Hawking's theorem are the existence of a homeomorphism and the preservation of future directed null geodesics by f and f^{-1}, both of which imply a C^r diffeomorphism for $r \geq 3$, Malament then went onto generalising these results to future and past distinguishing spacetimes which, as the reader will recall, is a weaker causality condition than strong causality.

Theorem 2.3 (Malament Theorem 1) *If $f : (M, g) \rightarrow (M', g')$ is a bijection which preserves future directed continuous timelike curves, then f is a homeomorphism.*

Theorem 2.4 (Malament Theorem 2) *If $f : (M, g) \rightarrow (M', g')$ is a chronological bijection between future and past distinguishing spacetimes, then f is a homeomorphism.*

It follows from Theorem 2.2 that f is a smooth conformal isomorphism.

Finally, a result from Kronheimer and Penrose [1] shows that a chronological bijection between distinguishing spacetimes is also a causal bijection. We summarise these HKMMKP theorems into the single theorem:

Theorem 2.5 HKMMKP Theorem: *A causal bijection between two future and past distinguishing C^r n-dimensional spacetimes for $r \geq 3$ is a C^r conformal isometry.*

As in the case of strongly causal spacetimes, the HKMMP theorem shows that, for future and past distinguishing spacetimes, the differentiable structure of a spacetime can be obtained from the chronological structure. In future and past distinguishing spacetimes, strong causality is violated on sets of zero measure, along inextendible null geodesics [83] about which the map between the Alexandrov topology and the manifold topology breaks down (see Fig. 2.4). In [91] this peculiar feature was used to show that causal bijections between two future and past distinguishing spacetimes can only exist when their dimensions are the same. This is also implicit in the Malament theorems.

The HKMMKP theorem therefore leads to a profound change in our perspective on spacetime (M, g). Instead of being an odd modification of Riemannian space, the Lorentzian spacetime "body" (M, g) can be thought to be made up of two ingredients: (a) its causal structure $(M, \prec)$, as its "skeleton" and (b) its conformal factor (equivalently its local volume element ϵ), as its "flesh". Thus, for a future and past distinguishing spacetime (M, g), one can summarise this theorem as

$$(M, g) = (M, \prec) + \epsilon. \tag{2.12}$$

This implies that *all* topological and geometric features of a future and past distinguishing spacetime can be obtained from the causal structure and the local volume element.

Echoing the Kronheimer-Penrose quote, this decomposition suggests a discretisation of the causal structure as a "first quantisation" of quantum gravity. As we will now describe, this is the route that CST takes.

2.3 The Causal Set Hypothesis

At the heart of any approach to quantisation is a choice of fundamental degrees of freedom, or ontology of quantum gravity. This in turn is informed by an identification of a quintessential aspect of continuum spacetime. In canonical approaches, for example, spacetime is viewed fundamentally in terms of the initial value problem, and hence the ontology is that of a constrained phase space. This constrained Hamiltonian system must then be quantised, for example, using the Dirac procedure. In continuum path integral approaches, it is the Lorentzian geometry which is taken as the fundamental field, and the focus is on evaluating expectation values of covariant observables.

What the HKMMKP theorem provides is a very different ontology of spacetime, one that places the spacetime causal structure $(M, \prec)$ centerstage. Mathematically, $(M, \prec)$ is a poset and this provides a unique starting point for quantisation. In CST, $(M, \prec)$ is "first quantised", or discretised, by requiring that every finite volume Alexandrov interval in the continuum spacetime (M, g) contains only a finite number of "fundamental spacetime atoms", *not* an uncountable infinity of events (see Fig. 2.5.) Thus, a finite *number* is associated with a finite spacetime volume. The discretised causal structure thus intrinsically contains the missing ingredient in the HKMMKP theorem, namely the volume element.

Technically, the discretisation renders the continuum causal structure into a **locally finite poset**, or **causal set**, defined as follows.

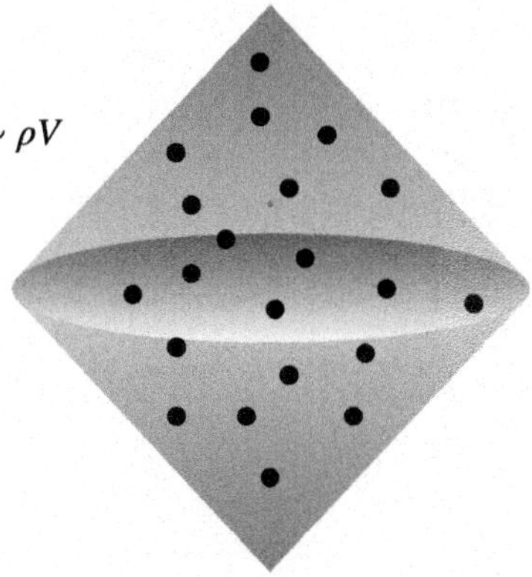

Fig. 2.5 Any finite spacetime volumes V contains a finite number of causal set elements n, where $n \sim \rho V$, and ρ^{-1} is the discreteness scale

Definition 2.1 A **causal set** C is a set with an order relation $\prec$ which, for $e, e' \in C$, is:

1. **Acyclic**: $e \prec e' \Rightarrow e' \nprec e$
2. **Transitive**: $e \prec e'$ and $e' \prec e'' \Rightarrow e \prec e''$, and
3. **Locally finite**: $|\mathrm{Fut}(e) \cap \mathrm{Past}(e')| < \infty$

where $|.|$ denotes the cardinality of the set and

$$\mathrm{Fut}(e) \equiv \{e' \in C | e \prec e',\ e' \neq e\}$$
$$\mathrm{Past}(e) \equiv \{e' \in C | e' \prec e,\ e' \neq e\}$$

We adopt the **irreflexive condition**, $e \nprec e$, and use the notation $\preceq$ when required, which allows $e \preceq e$.

When drawing a causal set, it is customary to show only the nearest neighbour relations or **links** in its **Hasse diagram** since the remaining relations can be deduced by transitivity. In Fig. 2.6 we show both the Hasse diagram of a causal set as well as the diagram with all the relations drawn in. The advantage of the former is obvious.

The CST hypothesis of Bombelli, Lee, Meyer and Sorkin [2] can be summarised as follows.

1. Quantum Gravity is a quantum theory of causal sets.
2. A causal set C is said to have a **continuum approximation** (M, g), if there exists an injective map $\Phi : C \hookrightarrow_\rho (M, g)$ such that (a) the order relation in C corresponds to the induced causal order in $\Phi(C)$ and (b) the cardinality of an order interval $(e, e') \subset C$ corresponds, in an appropriate sense, to the spacetime volume of the Alexandrov interval $I(\Phi(e), \Phi(e'))$ in units of the discreteness scale.

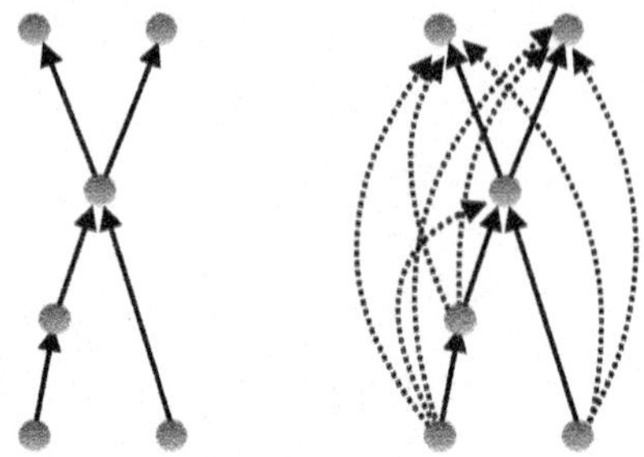

Fig. 2.6 The **Hasse diagram** of a 6 element causal set is shown on the left. Only the nearest neighbour relations or **links** are depicted. On the right is the same causal set with all the relations shown. Since all non-link relations can be deduced from transitivity, the Hasse diagram proves to be less messy while depicting causal sets

The first statement of the hypothesis is very general and does not refer to the continuum, or even the spacetime dimension. In it are hidden a host of questions: what does it mean to quantise causal sets, how should we interpret the results of the theory in terms of observables, etc. Versions of these questions are echoed in other approaches to quantum gravity, and the attempts to answer them are themselves typically approach dependent. The choice of *what* to quantise is therefore just the first step in a long, still unfinished journey in quantum gravity.

Whatever the dynamics of causal sets, however, one still needs to know when and how the continuum and classicality emerges in the theory. In particular, what does it mean for a causal set to look like spacetime? As with any fundamentally discrete theory of quantum gravity, CST needs a continuum approximation, since the discrete substructure must also contain all the necessary macroscopic information about spacetime. The second statement of the hypothesis thus helps "ground" causal set theory squarely inside Lorentzian geometry using a discrete version of the HKMMKP theorem, summarised by the CST slogan:

$$\text{Order} + \text{Number} \sim \text{Spacetime} \tag{2.13}$$

with the discrete-continuum correspondence:

$$\text{Order} \sim \text{Spacetime Causality} \tag{2.14}$$

$$\text{Number} \sim \text{Spacetime Volume} \tag{2.15}$$

In Sect. 2.3.2 we flesh out the precise nature of this correspondence, which defines the continuum approximation of the theory.

CST is not the first to identify the causal structure poset $(M, \prec)$ as the starting point for spacetime quantisation. Examples of poset approaches to quantum gravity are given in [1, 15, 81], and more recently in [19]. What distinguishes CST from these approaches is the specific nature of the continuum approximation using the discrete version of the HKMMKP theorem. The exception is Myrheim's seminal preprint [18] which contains many of the main ideas of CST.

2.3.1 Notation

We will use C to denote a causal set, $e \in C$ for an element of C and $\{e\}$ for the set of all elements of C. A **labelling** $L(C)$ of C is an injective assignment $\{e\} \to \mathbb{N}$. By convention, if $n = |C|$, then $L(C) : \{e\} \to (0, \ldots, n-1)$ or $L(C) : \{e\} \to (1, \ldots, n)$. The latter notation is preferred in this book, but both choices appear in the literature. We denote a **labelled causal set** by $\widetilde{C}$ and its elements by $\{e_1, e_2, \ldots e_n\}$. C is often referred to as an **unlabelled causal set** and $\widetilde{C}$ its labelled counterpart.

The labelling of the elements of a causal set is equivalent to a choice of coordinate system and hence any physical property of a causal set should be label independent.

The future and past sets of an element in C have been defined in Eq. (2.1). We define the **inclusive future** and **inclusive past** of a subset $X \subset C$

$$\mathrm{IFut}(X) \equiv X \cup \mathrm{Fut}(X), \quad \mathrm{IPast}(X) \equiv X \cup \mathrm{Past}(X). \tag{2.16}$$

An **order interval** $(e, e') \subset C$ is the subset

$$(e, e') \equiv \{e'' | e \prec e'' \prec e'\}. \tag{2.17}$$

and the **closed order interval**

$$[e, e'] \equiv \mathrm{IFut}(e) \cap \mathrm{IPast}(e'). \tag{2.18}$$

The set of **maximal**, or future-most elements, and the set of **minimal**, or past-most elements of C are defined as

$$\max(C) \equiv \{e \in C | \nexists e', e \prec e'\}, \quad \min(C) \equiv \{e \in C | \nexists e', e' \prec e\}. \tag{2.19}$$

Let $\widetilde{\Omega}_n$ denote the set of all labelled n-element causal sets and Ω_n the set of all unlabelled n-element causal sets. Let $\widetilde{C} \in \widetilde{\Omega}_n$ be a labelling of $C \in \Omega_n$. There is a natural action of the permutation group S_n on the elements of $\widetilde{C}$, i.e., for $\pi \in S_n$

$$\pi \circ \{e_1, e_2, \ldots e_n\} = \{e_{\pi(1)}, e_{\pi(2)}, \ldots e_{\pi(n)}\}. \tag{2.20}$$

If the order relations are kept unchanged, i.e., $e_{\pi(i)} \prec e_{\pi(j)} \Leftrightarrow e_i \prec e_j$, this produces another labelled causal set $\widetilde{C}' = L'(C)$ which is another labelling of C or a **relabelling** of $\widetilde{C}$.

However, not every relabelling yields a different labelled causal set . If $e, e' \in C$ have the same future and past sets, i.e., $\mathrm{Past}(e) = \mathrm{Past}(e')$ and $\mathrm{Fut}(e) = \mathrm{Fut}(e')$, then they are causally indistinguishable from each other. Permuting the labels on e and e' will therefore not produce a different labelling of C. The set of such redundant labellings is called the **automorphism group** $Aut(C)$ of C. $Aut(C)$ is clearly not the same for every causal set. Figure 2.7 shows simple examples of this.

The permutation group S_n therefore acts on $\widetilde{\Omega}_n$, but because of the non-triviality of the automorphism group, the action is not free, i.e., it has a non-trivial stabiliser subgroup, which moreover, varies over the different orbits in $\widetilde{\Omega}_n$. If $\widetilde{C} \sim \widetilde{C}'$ denotes equivalence under relabelling, every $C \in \Omega$ can be thought of as an equivalence class, $C = [\widetilde{C}]$.

An important class of labelled causal sets are those that are **naturally labelled**,

$$NL : \{e\} \to \mathbb{N}, \quad e_i \prec e_j \Rightarrow i < j. \tag{2.21}$$

The labels $i \in \mathbb{N}$ thus act like a "coordinate time" label on the elements. Natural labellings are also referred to as **linear extensions** in the literature. We will use this crucially in Sect. 5.3.1 for constructing the sequential growth dynamics of causal sets.

An **order invariant** over a field $\mathcal{F}$ is a map

$$\mathcal{O} : \Omega \to \mathcal{F}, \tag{2.22}$$

which is therefore label independent. Equivalently, it is a label-independent map

$$\mathcal{O} : \widetilde{\Omega} \to \mathcal{F}, \quad \mathcal{O}(\widetilde{C}) = \mathcal{O}(\widetilde{C}'), \; \forall \widetilde{C} \sim \widetilde{C}'. \tag{2.23}$$

An important and simple example of an order invariant is the number of relations in a causal set. More examples will be given in Chap. 3.

Two elements $e, e' \in C$ are said to be **linked** if $e \prec e'$ and $\nexists\, e'' \in C$ such that $e \prec e'' \prec e'$. Note that we are using the irreflexive condition, i.e., $e \nprec e$. We will denote a link by the symbol $\prec\!*$. In the Hasse diagram of a causal set (Fig. 2.6) only the links are shown, with the relations being induced via transitivity.

An m-**chain** c_m in C is a totally ordered subset of C (see Fig. 2.7), i.e., for every pair $e, e' \in \mathrm{c}_m$, either $e \prec e'$ or $e' \prec e$. Thus it admits a labelling such that $e_1 \prec e_2 \prec e_3 \ldots \prec e_m$. We denote by $\mathrm{c}(e, e')$ a chain starting from e and ending at e'. A chain made up only of link relations is called a **linked-chain** or **linked-path**.

An m-**antichain**, a_m is a collection of m mutually unrelated elements in C, i.e., for every pair $e, e' \in \mathrm{a}_m$, $e || e'$.

Since $\prec$ is irreflexive in C, e and e' do not belong to (e, e'). Typically we will also assume that $e, e' \in C$.

Exercise 2.2 Show that a labelled chain or a linked-path in an n-element causal set is not itself label invariant. Find an order invariant using the set of chains and linked-paths in a causal set.

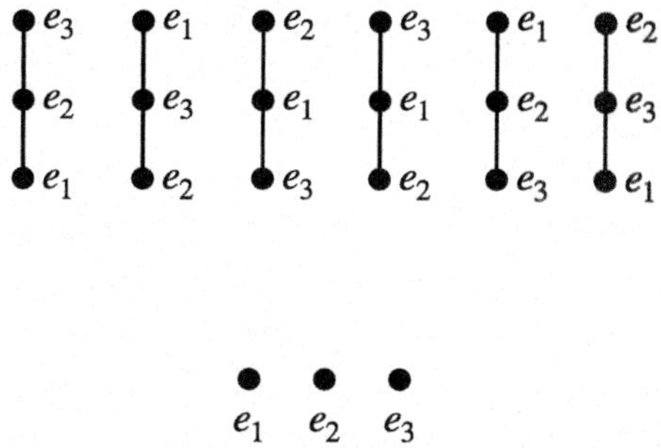

Fig. 2.7 Examples of Automorphisms. For the n-**chain** c_n which is a totally ordered set on n-elements, every possible relabelling produces a distinct labelled causal set. As shown, for c_3 there are 3! distinct relabellings, and hence $Aut(c_n)$ is trivial. On the other hand, for the n-**antichain** a_n which is a totally unordered set, every relabelling produces exactly the same labeled causal set. Hence $Aut(a_n) = S_n$

2.3.2 The Continuum Approximation

The local finiteness of causal sets means that discreteness in CST is regarded as fundamental and not just a tool for regulating the continuum. It is the continuum that is thought to be emergent, similar to a macroscopic fluid whose fundamental description is molecular, or discrete. Thus, in CST the **continuum approximation** is physical, but not the continuum limit. The latter of course is a useful "sanity check".

Implementing the discrete-continuum correspondence encoded in Eqs. (2.14) and (2.15) consistently is non-trivial and leads us to a distinctive formulation of the continuum approximation. In particular, standard discrete methods, commonly used in regular lattices or simplicial decompositions, are not available to CST. One is thus forced to find new tools and think in new ways about spacetime discreteness. We urge the reader to follow this section carefully, since it motivates and provides the basis for many aspects of CST.

We would like to determine the conditions under which a causal set C has a continuum approximation (M, g), which we denote by $C \sim (M, g)$.

To begin with, the order relation $\prec$ in C should correspond to the causal order $\prec_g$ in the approximating spacetime (M, g). This means that there exists a **causal embedding** which is an injective map $\varphi : C \hookrightarrow_\rho (M, g)$ such that for $e, e' \in C$, $e \prec e' \Leftrightarrow \varphi(e) \prec_g \varphi(e')$.

Exercise 2.3 Show that the 6-element crown poset $\mathbf{cr}_6$ (see Fig. A.5 in Appendix A) cannot be causally embedded into $\mathbb{M}^2$, but can be causally embedded into the cylinder spacetime $\mathbb{S}^1 \times \mathbb{R}$ with line element $ds^2 = -dt^2 + d\theta^2$ and hence also for $\mathbb{M}^d$, for $d > 2$.

Figure 2.8 shows a lightcone lattice $\mathcal{L}^2$ in $\mathbb{M}^2$, whose elements lie at integer values of the light cone coordinates $(\mathbf{u}, \mathbf{v})$, where $\mathbf{u} = \frac{1}{\sqrt{2}}(t - x)$ and $\mathbf{v} = \frac{1}{\sqrt{2}}(t + x)$. The edges of $\mathcal{L}^2$ are the nearest neighbour horismotic relations $(\mathbf{u}, \mathbf{v}) \rightarrow (\mathbf{u} + 1, \mathbf{v})$ and

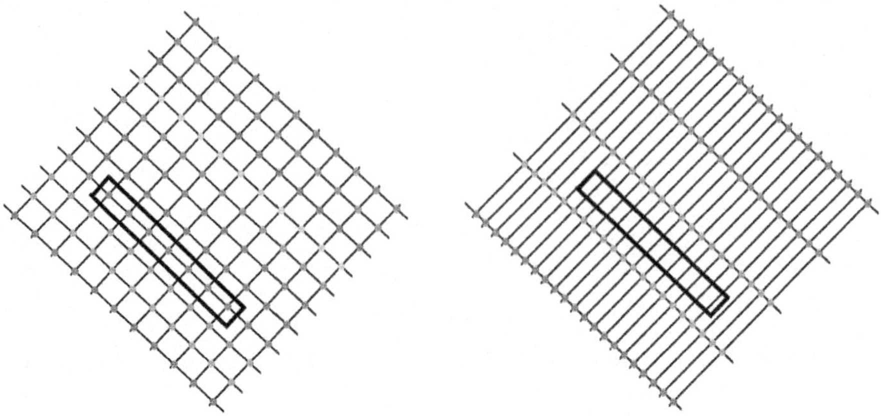

Fig. 2.8 In a regular lightcone lattice, the number to volume correspondence cannot be made in a covariant manner. On the left, an Alexandrov interval of volume 7 unit^2 contains exactly 7 elements. On the right, under a boost an Alexandrov interval of the same volume contains no elements at all

$(\mathbf{u}, \mathbf{v}) \to (\mathbf{u}, \mathbf{v}+1)$, with transitivity determining the remaining causal relations $\prec$. Thus $\mathcal{L}^2$ is a causal set that is causally embeddable in $\mathbb{M}^2$ (and therefore $\mathbb{M}^d$ for any $d \geq 2$.)

In this frame, there appears to be an exact number to volume correspondence: every four unit volume interval contains a single element. Under a boost $\beta = v/c$ about the origin, $\mathbf{u} \to \zeta\mathbf{u}$ and $\mathbf{v} \to \zeta^{-1}\mathbf{v}$, where $\zeta = \sqrt{(1+\beta)/(1-\beta)}$. If $\zeta = 2$, then, as illustrated in Fig. 2.8, the same four unit volume contains no element after the boost. Thus, the number-volume correspondence is violated, even for large volumes, and cannot be implemented in a covariantly consistent way. It is relatively straightforward to see that this is also true for any regular lattice. Since the spacetime geometry g is covariantly defined, a frame-dependent number to volume correspondence is therefore insufficient to implement the discrete version of the HKMMKP theorem covariantly. Without this, there can be no well-defined continuum approximation.

The quandary is partially resolved if instead we consider causal sets generated from the spacetime via a random point process which is uniform with respect to the spacetime volume. The idea of such a "random lattice" was suggested in the Euclidean context in [92] in order to preserve continuum symmetries. To make this precise we will assume that (M, g) is a **causally convex** spacetime region, i.e., for any $p, q \in M$, $I(p, q) \subset M$, and is of finite volume.

A **point process** on (M, g) is a random process whose values are a set of points on (M, g). In order for these points to satisfy a number to volume correspondence, they should be uniformily distributed with respect to the spacetime volume. A natural choice for a uniform point process is the **general Poisson point process (GPPP)** $\Phi_\rho(M, g)$ [93] for which:

1. The probability that there are n number of points in the spacetime region of volume V is

$$P_V(n) = \frac{(\rho V)^n}{n!} \exp^{-\rho V}, \tag{2.24}$$

where ρ is the **density** of the point process with respect to the volume measure in (M, g) (equivalently ρ^{-1} is the spacetime **discreteness scale**.) The number of points in a spacetime region of volume V is therefore a Poisson random number $\mathbf{n}$ such that

$$\langle \mathbf{n} \rangle = \rho V, \quad \Delta \mathbf{n} = \sqrt{\rho V}. \tag{2.25}$$

2. The number of points in k disjoint regions of (M, g) form k independent random variables.

Let $\sigma = \{\mathbf{x}_1, \mathbf{x}_2 \ldots \mathbf{x}_k\}$ denote a given realisation of $\Phi_{\underset{\sim}{\rho}}(M, g)$, where $\mathbf{x}_i$ are the coordinates of the ith point in (M, g). A **labelled causal set** $\widetilde{C}$ can be constructed from σ by identifying every $\mathbf{x}_i$ with an element $e_i \in \widetilde{C}$ and using the induced spacetime causality relations as follows: for every pair $\mathbf{x}_i, \mathbf{x}_j$ in σ, $\mathbf{x}_i \prec_g \mathbf{x}_j \Leftrightarrow e_i \prec e_j$. Let $\widetilde{\mathfrak{C}}_\rho(M, g)$ denote the **ensemble of labelled causal sets** obtained from $\Phi_\rho(M, g)$.

The permutation group S_k has a natural action on σ:

$$\pi \circ \{\mathbf{x}_1, \mathbf{x}_2 \ldots \mathbf{x}_k\} = \{\mathbf{x}_{\pi(1)}, \mathbf{x}_{\pi(2)} \ldots \mathbf{x}_{\pi(k)}\}, \tag{2.26}$$

where $\pi \in S_k$. However, not every permutation produces a distinct labelled causal set—a permutation between two elements e_i, e_j which have the same elements in their future and past sets, will yield the same labelled causal set. Thus, for every labelled causal set $\widetilde{C}$ there is a **relabelling subgroup** $R_k(\widetilde{C}) \subset S_k$ whose action is free on $\widetilde{\mathfrak{C}}_{\underset{\sim}{\rho}}(M, g)$. This defines a relabelling equivalence relation $\widetilde{C} \sim \widetilde{C}'$ on $\widetilde{\Omega}(M, g)$, and $\widetilde{C}$ and $\widetilde{C}'$ are termed **order isomorphic**. An **unlabelled causal set** C is an equivalence class $[\widetilde{C}]$ under relabellings.

We denote the ensemble of unlabelled causal sets generated from $\Phi_\rho(M, g)$ by $\mathfrak{C}_\rho(M, g)$, which we interpret as the kinematical "microstates" of (M, g) with respect to the discreteness scale ρ^{-1}.

Conversely, given a causal set, we can ask what spacetime, if any, it corresponds to. A causal set C is said to have a **continuum approximation** (M, g), which we denote by $C \sim_\rho (M, g)$, if there exists a causal embedding $\varphi : C \hookrightarrow_\rho (M, g)$ such that the number of points of $\varphi(e)$ are Poisson distributed with respect to the spacetime volume, i.e., according to Eqs. (2.24) and (2.25). C is moreover said to be **faithfully embedded** into a spacetime (M, g) at density ρ if there exists such a causal embedding $\varphi : C \hookrightarrow_\rho (M, g)$.

In much of the CST literature the GPPP $\Phi_\rho(M, g)$ is referred to as a **Poisson sprinkling** and we will use both these terms in this book. The GPPP can also be conceptualised as a "sampling" of the continuum spacetime (M, g). Although this is a useful picture, its important to keep in mind that in CST, the continuum is not fundamental, but an approximation of the underlying ensemble $\mathfrak{C}_\rho(M, g)$.

The uniformity of $\Phi_\rho(M, g)$ via the number-volume correspondence has startlingly strong consequences, some of which depend on its measure theoretic underpinnings, which we now flesh out (see [94].)

While we are taught to think of Lorentzian spacetime in terms of a metric and hence "distances", the latter are not distances in the strict mathematical sense, since they do not minimise lengths, can vanish between distinct spacetime events, and do not satisfy the triangle inequality. Instead, it is more natural to think of a Lorentzian spacetime as a measure space, a property also shared by Riemannian spaces. Consider a region $X \subset M$ with volume

$$\mathrm{vol}_g(X) \equiv \int_X \sqrt{-g(\mathbf{x})} d\mathbf{x} > 0 \tag{2.27}$$

where $d\mathbf{x}$ is the standard **Lebesgue** measure on X. Such a region is therefore a **measurable set** in (M, g) with respect to the spacetime volume measure. The collection of measurable sets forms a so-called Borel σ-algebra $\mathfrak{S}_M$, which is closed under countable unions, intersections and complements. Together, this makes any Lorentzian spacetime a **measure space**, defined as the **measure triple** $(M, \mathfrak{S}_M, \mathrm{vol}_g)$. This is the approach taken in [94] and we follow this exposition below.

Consider a finite volume spacetime, $\mathrm{vol}_g(M) < \infty$. Let $\Sigma_k(M, g)$ denote the set of all $\sigma_k = \{\mathbf{x}_1, \mathbf{x}_2 \ldots \mathbf{x}_k\}$ in a k-element binomial point process where $\mathbf{x}_i \in M$ is an event in (M, g). This is an uncountably large space and hence the probability for a given σ_k to be realised is a set of measure zero. However, there is a natural sense in which the (Lebesgue) volume measure on (M, g) can be used to induce a measure on subsets of $\Sigma_k(M, g)$.

Let us start with $k = 1$ and pick $\sigma = \{\mathbf{x}\}$, where $\mathbf{x}$ is a particular event in M. The probability density for $\mathbf{x} \in (M, g)$ to be realised is

$$p(\mathbf{x}|(M, g)) = \frac{1}{\mathrm{vol}_g(M)} \sqrt{-g(\mathbf{x})}. \tag{2.28}$$

If we now consider the set $S_X^{(1)}$ of all $k = 1$ sprinklings $\sigma(\mathbf{x})$ into a measurable region $X \subset M$, (i.e., one with a well defined volume $\mathrm{vol}_g(X)$) this gives us a probability measure

$$\mu_1(S_X^{(1)}) = \frac{1}{\mathrm{vol}_g(M)} \int_{\mathbf{x} \in X} \sqrt{-g(\mathbf{x})} d\mathbf{x} = \frac{\mathrm{vol}_g(X)}{\mathrm{vol}_g(M)}. \tag{2.29}$$

Each measurable set in (M, g) thus gives us a measurable set in $\Sigma_1(M, g)$, and this generates a σ-algebra $\mathfrak{S}_1(M, g)$. Hence $(\Sigma_1(M, g), \mathfrak{S}_1(M, g), \mu_1)$ is a probability measure space since $\mu_1(\Sigma_1(M, g)) = 1$.

Consider more generally a set of k-events $\sigma_k = \{\mathbf{x}_1, \mathbf{x}_2, \ldots, \mathbf{x}_k\} \in \Sigma_k(M, g)$. Since each point is independently sprinkled, the probability density is

$$p(\sigma_k|(M, g)) = \prod_{i=1}^{k} p(\mathbf{x}_i|(M, g)) = \frac{1}{\mathrm{vol}_g(M)^k} \prod_{i=1}^{k} \sqrt{-g(\mathbf{x}_i)}. \tag{2.30}$$

Again, for the set $S_X^{(k)}$ of all k-element sprinklings into in a measurable region $X \subset M$ the probability measure is

$$\mu_k(S_X^{(k)}) = \frac{1}{\mathrm{vol}_g(M)^k} \prod_{i=1}^{k} \int_{\mathbf{x}_i \in X} \sqrt{-g(\mathbf{x}_i)} d\mathbf{x}_i = \left(\frac{\mathrm{vol}_g(X)}{\mathrm{vol}_g(M)}\right)^k \tag{2.31}$$

Thus for every k, the Borel σ-algebra $\mathfrak{S}_M$ over (M, g) generates a σ-algebra $\mathfrak{S}_k$ over $\Sigma_k(M, g)$, and the measure triple $(\Sigma_\rho(M, g), \mathfrak{S}_k, \mu_k)$.

It is an easy step to see how an n-element binomial point process in (M, g) leads to the Poisson distribution. The probability for there to be exactly k elements out of n total elements to lie in $X \subset M$ is therefore

$$\binom{n}{k} \left(\frac{\mathrm{vol}_g(X)}{\mathrm{vol}_g(M)}\right)^k \left(1 - \frac{\mathrm{vol}_g(X)}{\mathrm{vol}_g(M)}\right)^{n-k}, \tag{2.32}$$

which, in the limit of large n, $\mathrm{vol}_g(M)$ with $\rho = n/\mathrm{vol}_g(M)$ fixed, gives the Poisson distribution Eq. (2.24) [94].

For a GPPP $\Phi_\rho(M, g)$, while $\langle k \rangle = \rho \mathrm{vol}_g(X)$, the number of elements k in a given realisation $\sigma = \{\mathbf{x}_1, \mathbf{x}_2, \ldots \mathbf{x}_k\}$ is not fixed and can lie any where in $\mathbb{N}$. Let (M, g) *not* be of finite volume, for example, $\mathbb{M}^d$. $\Phi_\rho(M, g)$ then generates the set $\Sigma_\rho(M, g)$ of countable strings $\sigma = \{\mathbf{x}_1, \mathbf{x}_2, \ldots\}$. For every measurable set $X \subset M$, define the **cylinder set** $\mathrm{cyl}(k, X)$ as the set of all $\sigma \in \Sigma_\rho(M, g)$ such that exactly k-elements of σ lie in X, with the rest lying anywhere else in M. Its probability measure is therefore

$$\mu_\rho(\mathrm{cyl}(k, X)) \equiv \frac{1}{k!} e^{-\rho \mathrm{vol}_g(X)} (\rho \mathrm{vol}_g(X))^k, \tag{2.33}$$

which is the Poisson measure. The set of all cylinder sets $\mathrm{cyl}(k, X)$ over all sets X in the Borel-σ algebra over M then generates the σ-algebra $\mathfrak{S}_\rho$ over $\Sigma_\rho(M, g)$. Hence $(\Sigma_\rho(M, g), \mathfrak{S}_\rho, \mu_\rho)$ is a probability measure space. We will use this in the proof of Lorentz invariance in $\mathbb{M}^d$ in Sect. 2.3.2.3.

Before we end this introduction to the continuum approximation, let us pause a moment to reflect on the choice of a GPPP. Since it is the limit of a binomial point process, where a single element is assigned to a unit volume in a uniform manner, the GPPP seems the natural choice to implement the number to volume correspondence. A pertinent question is whether a different choice of distribution might lead to a different manifestation of the continuum approximation. However, the most optimal choice of distribution is one in which the fluctuations are minimised. In [95] it was shown that for $d > 2$, and under certain further conditions, the GPPP indeed does the best job. While strengthening these results would be an interesting exercise, combined with the argument for naturalness, the GPPP leads to compelling physical predictions for CST. We will see in Sect. 2.3.2.3 that the choice of GPPP ensures a strong version of Lorentz invariance, and in Sect. 6.2, that the form of Δn (Eq. (2.25)) is key to Sorkin's accurate 1987 prediction of the fluctuation in the cosmological constant.

For this reason, we will henceforth define the continuum approximation in CST using the GPPP. We now examine the consequences of this choice.

2.3.2.1 Nonlocality

Figure 2.9 shows the Hasse diagram of a causal set C obtained by Poisson sprinkling into a causal diamond $\mathbb{D}^2 \subset \mathbb{M}^2$, or Alexandrov interval $I(p, q)$ of volume V, at density $50/V$. This messy graph has some striking features, which we now examine. We have already referred to the relation between nearest neighbours as **links**. Every element $e \in C$ thus has a future and a past **link set**

$$L^+(e) \equiv \{e' \in C | e \prec\!* \, e'\}, \quad L^-(e) \equiv \{e' \in C | e' \prec\!* \, e\}. \tag{2.34}$$

We notice in the causal set of Fig. 2.9 that the cardinality of $L^+(e)$ and $L^-(e)$ varies considerably over C—from a value of $|L^+(e)| = 7$ down to a value of 2, even in the lower part of the graph. This makes it very different from simplicial decompositions of spacetimes, where the dual graph has a fixed number of nearest neighbours or **valency**.

More generally, in a continuumlike causal set $C \sim_\rho (M, g)$, the cardinality of the future and past link sets of the elements, or equivalently their **valency**, varies from element to element for the following reason. Let $\varphi : C \hookrightarrow_\rho (M, g)$ be a faithful embedding and let $p = \varphi(e)$ for some $e \in C$. By a slight abuse of notation we will use $L^+(p)$ to denote the future link set of $p \in \varphi(C)$ in $\varphi(C)$. The probability for an event $q \in (M, g)$ to be in $\varphi(C)$ and also in $L^+(p)$ is

$$dV_q e^{-\rho \mathrm{v}} \tag{2.35}$$

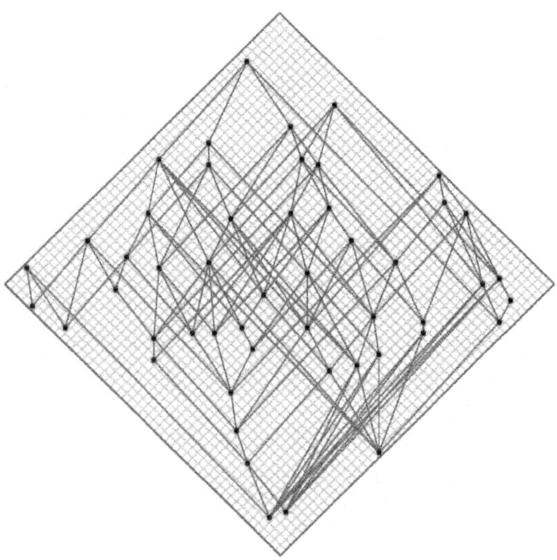

Fig. 2.9 The Hasse diagram of a causal set obtained from a Poisson sprinkling into $\mathbb{D}^2$ of volume V at density $50/V$

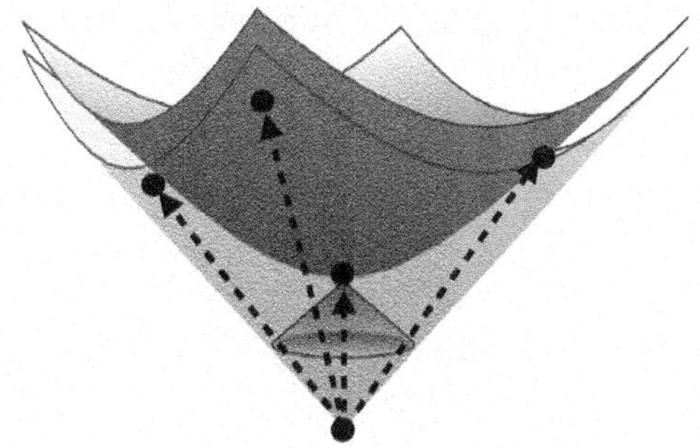

Fig. 2.10 For a causal set obtained from a Poisson sprinkling into $\mathbb{M}^2$, the number of links or nearest neighbours is infinite. The resulting graph is therefore of infinite valency

where $\mathbf{v} \equiv \mathrm{vol}(p,q)$, the volume of $I(p,q)$. This is the probability for there to be no elements of $\varphi(C)$ in $I(p,q)$, *given* $p,q \in \varphi(C)$. This is significant only for $\mathrm{vol}(p,q) \sim \rho^{-1}$ which determines the locus of the elements $L^+(p)$, upto fluctuations. In $\mathbb{M}^d$, for example, define the constant volume hyperboloid $\mathbb{H}^+(\rho^{-1}) \subset I^+(p)$ as the locus of points such that $\mathrm{vol}(p,r) = \rho^{-1}$ or equivalently, $\tau(p,r) = (\rho\zeta_d)^{-1/d}$ where ζ_d is a constant (see the Notations section.) Most of the elements in $L^+(p)$ therefore lie in a collar neighbourhood of $\mathbb{H}^+(\rho^{-1})$ in between the regions of volume $\rho^{-1} \pm \sqrt{\rho^{-1}}$ (Fig. 2.10.) This is a non-compact, infinite volume region and hence $|L^+(p)|$ is (almost surely) infinite. Thus, the cardinality of the set of nearest neighbours in the causal set graph, or its valency, is infinite. This is a quintessential feature of a continuumlike causal set.

Riemannian intuition tells us that locality corresponds to the compactness of the set of nearest neighbours. However in the Lorentzian case, the nearest neighbours belong to a non-compact set of infinite volume. In the continuum, the tangent space structure can be used to recover a semblance of locality. This is no longer the case in the fundamentally discrete construction of a causal set C where there is no intrinsic tangent space structure. This results in a characteristic **nonlocality** for continuumlike causal sets and makes its difficult presence known when reconstructing the continuum geometry, as we will see in Chap. 3, where standard lattice constructions are no longer meaningful.

2.3.2.2 Voids

While the average number of elements of a continuumlike causal set $C \sim_\rho (M,g)$ in any finite volume region $X \subset M$ is given by $\rho\mathrm{vol}_g(X)$, a striking feature of the GPPP is the possibility of large "voids" in the continuum spacetime, i.e., regions that contain *no* causal set elements. For an unbounded spacetime like $\mathbb{M}^d$, the probability for the existence of a void of *any* size is one, as is the existence of a large number of large voids!

The presence of a macroscopic void means that the continuum approximation breaks down in this region. Conversely, CST predicts that the emergent continuum regions of spacetime are bounded both spatially and temporally, even if the underlying causal set is itself countable. In other words, a continuum universe is not viable "forever". Of course, in the current phase of the observable universe, at least from the known early universe to the present, the continuum approximation should be valid. Thus, CST can be a viable theory only if the probability of occurrence of a macroscopic void anywhere in our observable universe can be shown to be negligible. In [96] the probability for there to be at least one nuclear size void $\sim 10^{-60}m^4$ was calculated in a region of Minkowski spacetime which is of the size of our present universe. Using general considerations, and assuming a Planck-scale cut-off $\rho = (1/\ell_p)^d$, it was shown to be of order $e^{-10^{72}}$, which is an absurdly small number! This implies that CST poses no phenomenological inconsistency in this regard. Nevertheless, the implications for a finite (but very large) continuum approximation is important to keep in mind.

Exercise 2.4 Use the calculation of [96], referred to above, to find out how old the universe will need to be for there be a void of dimension $10^{-20}m$ with probability 0.01.

2.3.2.3 Lorentz Invariance

Regular lattices, like the light cone lattice in $\mathbb{M}^2$, break the continuum symmetries of the spacetime or space from which they are generated, with the discrete symmetry group being a small subgroup of the continuous symmetry group. Hence we often hear the phrase, "the lattice breaks spacetime symmetries". In $\mathbb{M}^d$, any regular lattice breaks Lorentz invariance. However, Lorentz invariance has been experimentally verified to a high degree of precision [73,97], and this provides an important constraint to any theory of quantum gravity in which discreteness is considered to be fundamental. The question we must therefore ask is whether Lorentz invariance is broken by causal sets which are approximated by $\mathbb{M}^d$. Bombelli, Henson and Sorkin [72] showed that this is not the case, in *any* realisation of a causal set in $\mathfrak{C}_\rho(M, g)$.

As we showed earlier, the Poisson sprinklings into $\mathbb{M}^d$ forms a probability measure triple $(\Sigma_\rho, \mathfrak{S}_\rho, \mu_\rho)$. Let $\mathcal{L}$ denote the Lorentz group in $\mathbb{M}^d$. The action of any $\Lambda \in \mathcal{L}$ on $\mathbb{M}^d$ takes one realisation C of the GPPP to another one C', where $C, C' \in \Sigma_\rho$ and hence $\mathcal{L}$ also acts on $\widetilde{\mathfrak{C}}_\rho(\mathbb{M}^d)$. Since Lorentz transformations preserve spacetime volumes, this means that the probability measure space $(\Sigma_\rho, \mathfrak{S}_\rho, \mu_\rho)$ remains invariant

$$\mu_\rho = \mu_\rho \circ \Lambda, \quad \forall \Lambda \in \mathcal{L}. \tag{2.36}$$

Let $\mathbb{H}$ denote the hyperboloid of unit future-directed timelike vectors from the origin. To demonstrate Lorentz invariance we must show that a preferred direction vector cannot be *consistently* assigned to each C.

Assume otherwise, i.e., that for every $C \in \Omega_\rho$ we can pick a vector in $\mathbb{H}$, which is intrinsic to $\varphi(C)$. For example, fix an element $p \in \varphi(C)$ as the origin, and assign the direction vector $\mathbf{d} \in \mathbb{H}$ to be the vector from p to a particular $q \in L^+(p)$ in $\varphi(C)$. Since Lorentz tranformations act on $\mathfrak{C}_\rho(\mathbb{M}^d)$, consider $\Lambda \in \mathcal{L}$ about p, and let $C' = \Lambda \circ C$ with $p \to p$ and $q \to q' \neq q$. Consistency means that the direction vector $\mathbf{d}'$ in C' is the vector from p to q'.

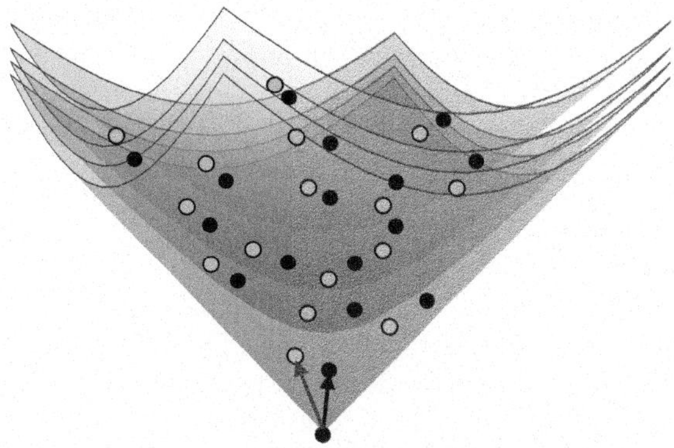

Fig. 2.11 The **a** black and **b** circled gray elements depict two different Poisson sprinklings C_a and C_b in $\mathbb{M}^d$ related by a boost about an element p, i.e., there is an induced action $\Lambda^* : C_a \to C_b$. If a direction vector $\mathbf{d}_a$ from p to its nearest neighbour in C_a is used to pick a preferred frame, **equivariance** implies that the preferred direction in C_b is $\mathbf{d}_b = \Lambda(\mathbf{d}_a)$

If $\mathbf{D} : \Omega_\rho \to \mathbb{H}$, then this means that $\mathbf{D}$ transforms **equivariantly** under Lorentz transformations

$$\Lambda \circ \mathbf{D} = \mathbf{D} \circ \Lambda. \tag{2.37}$$

$\mathbf{D}$ is surjective, since one can get from any $\mathbf{d} \in \mathbb{H}$ to any other via a Lorentz tranformation (Fig. 2.11).

Assume further, that there exists a sigma algebra $\mathfrak{S}_\mathbb{H}$ over $\mathbb{H}$ such that $\mathbf{D}$ is a measurable map, i.e., for every $\alpha \in \mathfrak{S}_\mathbb{H}$, $\mathbf{D}^{-1}(\alpha) \in \mathfrak{S}_\rho$.[4] This in turn induces a probability measure

$$\mu_\mathbf{D} = \mu \circ \mathbf{D}^{-1} \tag{2.38}$$

over $\mathfrak{S}_\mathbb{H}$. Equations (2.36) and (2.37) imply that $\mu_\mathbf{D} = \mu_\mathbf{D} \circ \Lambda$, which leads to the following contradiction. For any $\alpha \subset \mathbb{H}$ such that $\alpha \in \mathfrak{S}_\mathbb{H}$ it is possible to choose a Λ such that $\alpha' = \Lambda \circ \alpha \subset \mathbb{H}$ is disjoint from α. Using $\mu_\mathbf{D}(\alpha) = \mu_\mathbf{D}(\alpha')$ therefore $\mu_\mathbf{D}(\alpha \sqcup \alpha') = 2\mu_\mathbf{D}(\alpha)$. By repeatedly acting on α with Λ, therefore, the union $\mu_\mathbf{D}(\bigsqcup_{k=0}^K \Lambda^k \circ \alpha) = K\mu_\mathbf{D}(\alpha)$, where K has no upper bound since $\mathbb{H}$ is non-compact. Since $\mu_\mathbf{D}(\mathbb{H}) = 1$, i.e., it is a probability measure, this is a contradiction.

Thus, there exists no measurable equivariant map $\mathbf{D}$.

Let us contrast this with the GPPP Φ_ρ into Euclidean space $\mathbb{R}^d$. This is also a probability measure space and the symmetries form the Euclidean group generated by $SO(d-1)$ rotations and translations, which leaves the measure invariant. However, unlike the Lorentzian case, it is possible to assign a direction vector for a given realisation. For example, we can take the direction vector to lie along the vector joining a given p in a realisation, and its nearest neighbour q, also in the same realisation. This defines an equivariant map $\mathbf{D} : \Omega_\rho \to \mathbb{S}^{d-1}$ with respect to the Euclidean group. It is moreover measurable since $\mathbb{S}^{d-1}$ is compact: the parameter K in the Lorentzian proof *is* bounded from above in this case, and the induced measure on $\mathbb{S}^{d-1}$ is therefore a probability measure. However, as one averages over the ensemble of point processes, there is no preferred direction.

---

[4] In measure theory a measurable map between two measure spaces has the same role as a continuous map between two topological spaces. It is therefore natural to assume that $\mathbf{D}$ is measurable.

This leads us to the conclusion that it is the non-compact nature of the Lorentzian space of directions $\mathbb{H}$ which implies the absence of a preferred direction even in a *single* realisation in $\mathfrak{C}_\rho(\mathbb{M}^d)$. In constrast to the Euclidean case, there is no possible choice of a "nearest neighbour" for the GPPP in $\mathbb{M}^d$, since there is no lower bound on the proper time, again because of the non-compactness of $\mathbb{H}$.

2.3.2.4 Coarse Graining

For any finite ρ, a given continuum-like C will faithfully embed into not just one but entire families of spacetimes. In other words, the ensembles $\mathfrak{C}_\rho(M, g)$ overlap. However, as ρ increases, one expects the overlaps to decrease. For a given spacetime (M, g), one can extract a one-parameter family of ensembles $\mathfrak{C}_\rho(M, g)$ by varying ρ. The continuum limit is obtained when $\rho \to \infty$. Conversely, as ρ decreases, the approximating spacetime becomes less reliable. A causal set $C \sim_\rho (M, g)$ can be **coarse grained** to a sub-causal set C' by using the point process at a density $\rho' < \rho$ to select elements of C. The resulting $C' \sim_{\rho'} (M, g)$, i.e., it belongs to $\mathfrak{C}_{\rho'}(M, g)$.

Consider a spacetime (M, g) with topological or geometric "structure" at scales $\ll \rho^{-1}$. A causal set $C \sim_\rho (M, g)$ will therefore not "pick up" this structure. As an example consider the flat, "Kaluza-Klein" spacetime $\mathbb{M}^d \times \mathbb{S}^1$, where the size of the compact direction is $\gg \rho^{-1/(d+1)}$. Then if $C \sim_\rho \mathbb{M}^d \times \mathbb{S}^1$, C is not approximated by $\mathbb{M}^d$. On the other hand, if the size of the compact direction is $\ll \rho^{-1/(d+1)}$, $C \sim_\rho \mathbb{M}^d \times \mathbb{S}^1$ and $C \sim_\rho \mathbb{M}^d$, i.e., the causal set doesn't probe the $\mathbb{S}^1$. Thus, we see that while coarse graining a larger causal set into a smaller one means losing some continuum information, it also brings out emergent features.

Using a different type of "decimation" or coarse graining, it is possible to obtain continuumlike causal sets from non-continuum like causal sets. For example, a regular lattice-like structure can be attached to each element of a continuumlike causal set, similar to a Kaluza–Klein fibration. Such a causal set is not itself continuumlike. Instead of the random coarse graining procedure, the original continuumlike causal set can be recovered by taking the quotient with respect to this fibre. Thus, even if the deep quantum regime in CST is not continuumlike, there could be a coarse grained, emergent continuum approximation.

2.3.3 The Hauptvermutung of CST

As we've discussed above, given a continuum approximation $C \sim_\rho (M, g)$, there exist other spacetimes (M', g') such that $C \sim_\rho (M, g)$, i.e., the continuum approximation is not unique for a given density ρ. The example of a causal set C being approximated by Kaluza-Klein spacetime $\mathbb{M}^d \times \mathbb{S}^1$ shows that if the size of the $\mathbb{S}^1$ is $\ll \rho^{-1/(d+1)}$ then the same causal set also approximates $\mathbb{M}^d$. Thus the continuum approximations can differ even in dimension, when the extra dimensions are physically irrelevant. The non-uniqueness is dependent on ρ and hence one expects that it goes away in the continuum limit $\rho \to \infty$ [98], which of course is not of physical interest to CST.

Since the spacetime *is* the approximation and not the causal set, one way to interpret this non-uniqueness is that the differences in the two spacetime are not relevant. Going back to the fluid analogy, while we can mathematically allow for structure, say turbulence, at scales much smaller than the characteristic scale below which the fluid description breaks down, these solutions do not carry any physical significance. From a fundamentally discrete view of spacetime this is also true of continuum spacetime structure below the discreteness scale. Non-uniqueness of the continuum approximation below the discreteness scale ρ^{-1} is therefore physically allowed in CST. On the other hand, there would be a serious breakdown in the continuum approximation if a causal set could have continuum approximations which differ at scales much larger than ρ^{-1}, and would mean that there is a problem with the discrete CST version of the HKMMKP theorem. We state this as the fundamental conjecture or Hauptvermutung of CST:

The Hauptvermutung of CST: *$C \sim_\rho (M, g)$ and $C \sim_\rho (M', g')$ implies that $(M, g) \approx_\rho (M', g')$, i.e., they are approximately isometric.*

Another way of stating this is that if the GPPPs for the two spacetimes (M, g) and (M', g') for a given density ρ are identical (as distributions), then these two spacetimes are the same "upto" scales of order ρ^{-1}.

Why should the Hauptvermutung hold? Based on an unpublished result by Meyer and Sorkin, Bombelli and Meyer [98] sketched a proof showing that in the limit $\rho \to \infty$ the GPPP for a given spacetime (M, g) converges (uniquely) to that spacetime. Later, Panangaden and Martin showed that globally hyperbolic spacetimes can be reconstructed entirely from countable causal sets [99]. More recently, Braun showed that for any future and past distinguishing spacetimes (that are also complete in a specific sense) the equivalence of the probability distribution of the underlying *labelled* random causal sets at all densities, implies a smooth isometry [100].

For finite ρ, there is no result on approximate isometries. Let us assume otherwise, i.e., that the same GPPP can be identified with two *macroscopically* distinct spacetimes at density ρ. This could, for example, suggest a hidden duality between these spacetimes, as is the case for mirror manifolds in string theory [101]. Such a dual description however is contrary to the spirit of the continuum approximation of CST where the idea is not to get rid of the fundamental ontology of spacetime itself, as in string theory, but to retain only specific aspects of it. Such a dual description however is contrary to the spirit of the continuum approximation of CST where the idea is not to get rid of the fundamental ontology of spacetime itself, as in string theory, but to retain only specific aspects of it. A causal set $C \sim_\rho (M, g)$ is therefore supposed to contains *all* the information in (M, g) at scales much larger than ρ^{-1}, i.e., it is a discrete version of the HKMMKP theorem. This is not compatible with macroscopic non-uniqueness. Although perhaps not the most pressing of questions in quantum gravity, finding a proof for the Hauptvermutung is important in making CST a complete and consistent theory.

The main technical hurdle for proving this theorem is in its statement itself, i.e., how to quantify what is meant by the term *approximate isometry* or, equivalently, how to define the closeness of spacetimes upto the scale ρ^{-1}. The closeness of Lorentzian geometries (and more general spaces) is a subject of current active research in Lorentzian geometry, some of it based on Lorentzian modifications of Gromov Hausdorff theory [94, 102–108]. A discussion of this is well outside the scope of this book—for most purposes it suffices to use the intuitive and functional understanding of closeness.

However, for the sake of pedagogy, we will briefly introduce a closeness criterion for Lorentzian spacetimes due to Bombelli which uses the GPPP in a critical way. Because of its simplicity it is simultaneously instructive for understanding and implementing the GPPP for CST. The key idea is to use the difference in the probabilities for getting the same causal set in the GPPPs for different spacetimes, to define a measure of their closeness.

Consider the class of spacetimes (M, g) with finite volume $\mathrm{vol}_g(M)$. Consider the binomial point process $\Phi_n(M, g)$ which generates the set $\Sigma_n(M, g)$ of n-element strings $\sigma_n = \{\mathbf{x}_1, \mathbf{x}_2, \ldots \mathbf{x}_n\}$. As we saw earlier, the Borel σ-algebra on (M, g) generates a σ-algebra $\mathfrak{S}_n$ over $\Sigma_n(M, g)$ with a probability measure μ_n given by Eq. (2.31), making it a measure space $(\Sigma_n(M, g), \mathfrak{S}_n, \mu_n)$. Associated to each σ_n, moreover, is a labelled causal set $\widetilde{C} \in \widetilde{\Omega}_n$ where the order relation is obtained from the induced causal relations of the events in (M, g). This defines the map $\Psi_g : \Sigma_n(M, g) \to \widetilde{\Omega}_n$, but it is (infinitely) many to one—by wiggling around the points in σ one can obtain a new $\sigma' \in \Sigma_n(M, g)$ which corresponds to the same labelled causal set $\widetilde{C}$. Figure 2.12 provides a simple illustration of this.

Let $\widetilde{\mathfrak{C}}_n \equiv \mathrm{Im}(\Psi_g)$ denote the ensemble of labelled causal sets that is generated in $\widetilde{\Omega}_n$ and $S_g(\widetilde{C}) \equiv \Psi_g^{-1}(\widetilde{C}) \subset \Sigma_n(M, g)$ denote the pre-image of $\widetilde{C} \in \widetilde{\mathfrak{C}}_n$ in $\Sigma_n(M, g)$. For every $\mathbf{x}_i \in \sigma$, let $\tilde{e}_i \equiv \Psi_g(\mathbf{x}_i)$ denote the corresponding element in $\widetilde{C}$. We define a probability measure on $S_g(\widetilde{C})$ as follows. The wiggle room for each $\mathbf{x}_i \in \sigma$, i.e., the set of all $\mathbf{x}_i'$ such that σ' remains in $S_g(\widetilde{C})$, is constrained since the relations between the $\mathbf{x}_i$ should not change. Working up from element $\mathbf{x}_1$ to $\mathbf{x}_n$, define

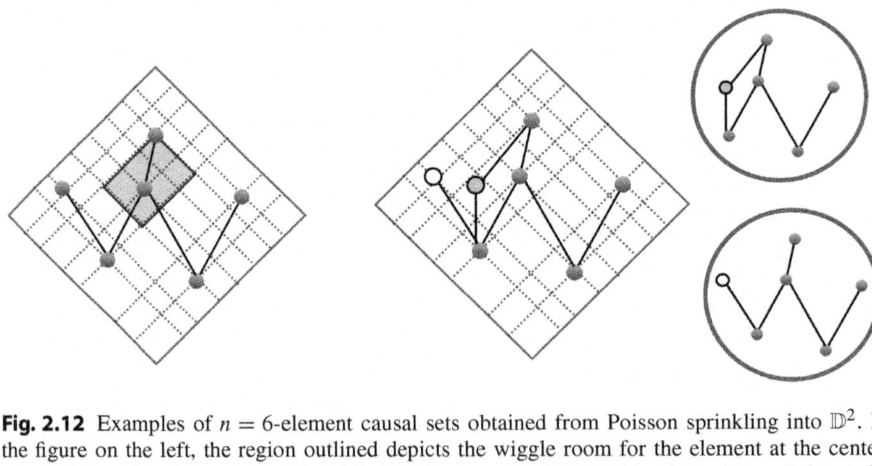

Fig. 2.12 Examples of $n = 6$-element causal sets obtained from Poisson sprinkling into $\mathbb{D}^2$. In the figure on the left, the region outlined depicts the wiggle room for the element at the center, when all the other elements are fixed. Moving the element within in this region does not change its relationship with the other elements and hence the causal set remains the same. The figure on the right shows an example of two causal sets that are equally likely to arise from an embedding into a region of $\mathbb{M}^2$. The circled grey element is replaced by the circled white element, which changes the induced relations

the region R_i as the wiggle room for the ith element, given the location of all the previous $i-1$ elements. Thus, the first element $\mathbf{x}_1$ has the freedom to lie anywhere in M, i.e., $R_1 = M$, while the next element $\mathbf{x}_2$ lies in (a) $R_2 = M \cap I^+(\mathbf{x}_1)$ if $\mathbf{x}_1 \prec_g \mathbf{x}_2$, or (b) $R_2 = M \cap I^-(\mathbf{x}_1)$ if $\mathbf{x}_2 \prec_g \mathbf{x}_1$, or (c) $R_2 = M \cap (I^+(\mathbf{x}_1) \cup I^-(\mathbf{x}_1))^c$ otherwise. Continuing this way, the measure on $S_g(\widetilde{C})$ is given by

$$\begin{aligned}\mu_n(S_g(\widetilde{C})) &\equiv \int_{S_g(\widetilde{C})} \widetilde{P}(x_1, \ldots x_n|(M,g))dx_1dx_2\ldots dx_n \\ &= \frac{1}{\mathrm{vol}(M)^n}\prod_{i=1}^n \int_{R_i}\sqrt{-g(x_i)}dx_i, \end{aligned} \tag{2.39}$$

where the R_i are obtained iteratively as described above (contrast this with the expression Eq. (2.31).) To get the probability for the unlabelled causal set $C \in \Omega_n$, one has to first permute over all the $n!$ labellings of C. However, as we noted in Sect. 2.3.1, not every relabelling yields a different labelled causal set. Thus one has to also quotient out by the automorphism group $Aut(C)$. The probability for any finite element causal set $C \in \Omega_n$ is therefore

$$p_n(C|(M,g)) \equiv \frac{n!}{|Aut(C)|}\frac{1}{\mathrm{vol}(M)^n}\prod_{i=1}^n \int_{R_i}\sqrt{-g(x_i)}dx_i. \tag{2.40}$$

Thus $p_n(C|(M,g))$ defines a probability measure $\Pi_n(M,g)$ on the finite space Ω_n, in which each C is measurable. Notice that for a causal set $C \in \Omega_n$ that does not embed into (M,g), at least one of the regions R_i must be of zero volume, and hence its probability is strictly zero. In general $\Pi_n(M,g)$ is complicated to calculate, not only because of the iteratively determined sets R_i, but also because $Aut(C)$ is not easy to characterise in general. Numerical methods could help, but again, as n becomes large the space of causal sets generated by the binomial point process also grows.

Exercise 2.5 Find the number of automorphisms for each labelled n-element causal set for $n = 2, 3, 4$, and plot the ratio of the number of n-element causal sets with non-trivial automorphisms to those with trivial automorphisms as a function of n. The set of all 2, 3 and 4-element causal sets are listed in Appendix A.

Each causal set $C \in \Omega_n$ can be thought of as a "probe" of the spacetime (M,g). As n becomes large, one expects that the probability distribution $p_n(C|(M,g))$ over Ω_n will typically be different for different (finite volume) spacetimes and hence can be used as a measure of their "closeness". There are many definitions of distance functions over probability spaces. Here we follow Bombelli's geometric formulation.

Since $\sum_{C\in\Omega_n} p_n(C|(M,g)) = 1$, $\Pi_n(M,g)$ is a point on the $d_n = |\Omega_n|$ dimension sphere $\mathbb{S}^{d_n} \subset \mathbb{R}^{d_n}$ with cartesian coordinates $\{p_n(C|(M,g))\}$.[5] We can therefore represent $\Pi_n(M,g)$ as a unit vector $|(M,g)\rangle$ in $\mathbb{R}^{d_n}$. The overlap of the probability distributions for two different spacetimes is given by the inner product

$$\langle (M_1, g_1)|(M_2, g_2)\rangle = \cos\theta_{12} \tag{2.41}$$

from which we define the closeness function

$$d_n((M_1, g_1), (M_2, g_2)) \equiv \theta_{12} = \frac{2}{\pi}\cos^{-1}\left(\sum_{C\in\Omega_n} \sqrt{p_n(C|(M_1, g_1))}\sqrt{p_n(C|(M_2, g_2))}\right). \tag{2.42}$$

When $g_1 = g_2$, the sum on the right hand side goes to 1 and hence $d_n(g,g) = 0$ as expected. But this is not a distance function since for finite n, it is still possible that that $|(M_1, g_1)\rangle = |(M_2, g_2)\rangle$, i.e., $d_n((M_1, g_1), (M_2, g_2)) = 0$. As an example consider a spacetime constructed from (M,g) by stitching in a region of negligible volume. Let $M' = M\#A$, where # denotes a topological connected sum and let $g'|_M = g$ and $\mathrm{vol}_{g'}(A) \ll \mathrm{vol}_g(M)/n$. The extra region A thus contributes negligibly and hence $|(M,g)\rangle = |(M', g')\rangle$ for finite n, therefore $d_n((M,g),(M',g')) \approx 0$.

We illustrate this construction with the simplest non-trivial case of $n = 2$, where $\Omega_2 = \{\bullet\bullet, \vdots\}$ with a_2 the the 2-element **antichain** $\bullet\bullet$ and c_2 the 2-element **chain** $\vdots$. For any (M,g), $P_2(\mathrm{a}_2|(M,g)) + P_2(\mathrm{c}_2|(M,g)) = 1$ which lies at the point $\theta(M,g) \equiv (\sqrt{P_2(\mathrm{a}_2|g)}, \sqrt{P_2(\mathrm{c}_2|g)})$ on the first quadrant of a unit circle $\mathbb{S}^1$ in $\mathbb{R}^2$ as shown in Fig. 2.13. Two distinct (M,g) and (M',g'), correspond to two different points $\theta(M,g)$ and $\theta(M',g')$ on $\mathbb{S}^1$. The arc distance between them can therefore be used to obtain a closeness function. Being a a unit circle, the closeness function is proportional to the angle subtended by the two points at the center of the circle

$$d_2(g, g') = \frac{2}{\pi}\cos^{-1}\left(\sum_{C\in\Omega_n} \sqrt{P_2(C|g)}\sqrt{P_2(C|g')}\right). \tag{2.43}$$

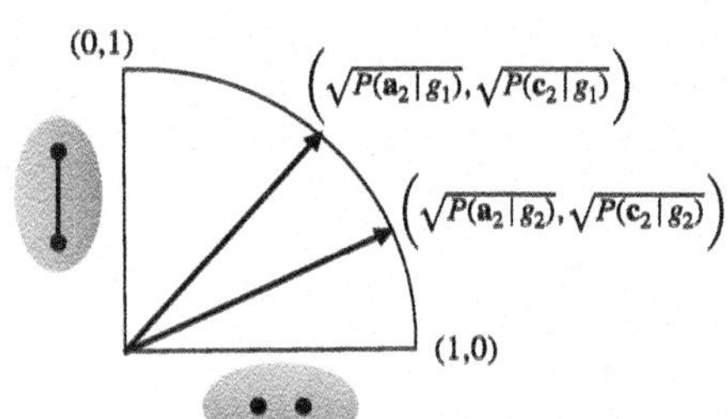

Fig. 2.13 For $n = 2$ the space of probabilities is just two dimensional. In the diagram, the probability for two different spacetimes g_1 and g_2 is shown

[5] As we will discuss in the next few chapters, d_n grows as $2^{\frac{n^2}{4}}$, which of course is very daunting. Nevertheless, the dominant contribution to this number is from causal sets that do not faithfully embed into any continuum spacetime and hence one expects that, effectively, d_n is much smaller.

Consider the following (unnatural) examples: Let (M, g) be the $0 + 1$ dimensional open spatial interval $(-1, 1)$ and (M', g') the $1 + 0$ dimension open time interval $(-1, 1)$. Clearly, $P_2(\mathrm{a}_2|(M, g)) = 1$ and $P_2(\mathrm{c}_2|(M, g)) = 0$, while $P_2(\mathrm{a}_2|(M', g'))$ $= 0$ and $P_2(\mathrm{c}_2|(M', g')) = 1$. The angle between these two points is $\pi/2$ and hence the distance between these spacetimes is the largest. Here, by reducing the dimensions, we have excluded the possibility that these spacetimes allow an embedding of one of the two 2-element causal sets. However, if we were to "thicken" these spacetimes by a finite amount, making them $1 + 1$ dimensional: $(M, g) = (-\epsilon, \epsilon) \times (-1, 1)$ and $(M', g') = (-1, 1) \times (-\epsilon', \epsilon')$, then the probabilities $P_2(\mathrm{c}_2|(M, g))$ and $P_2(\mathrm{a}_2|(M', g'))$ become non-zero, even for small but non-zero ϵ, ϵ'. In general, for finite n, the maximum distance is never realised since there is always a small overlap of the distributions between any pair of spacetimes.

As we noted, the closeness function $d_n((M, g), (M', g'))$ is not a distance function since at finite n it cannot distinguish between spacetimes which differ at scales smaller than ρ^{-1}. From the causal set point of view this is sufficient, but in order to define a true Lorentzian distance function using this construction, some more work needs to be done, a discussion of which is outside the scope of this book. The interested reader should look at [94] as well as the follow up of these ideas in [104] where a Lorentzian Gromov-Hausdorff distance function on the space of Lorentzian geometries is defined. A "null distance function" has also been used to define a Lorentzian distance function [109].

While there has been progress in defining various distance functions on the space of Lorentzian spacetimes, it is still insufficient to prove the Hauptvermutung in all its generality. As we will see in the next chapter, however, there is strong evidence in its favour from the process of geometric reconstruction. If an order invariant O is geometric, namely it can be shown to correspond to a spacetime covariant observable G, then this is equivalent to showing the following (weak) form of the fundamental theorem:

Theorem 2.6 ***O*-fundamental theorem** *If there exists a geometric order invariant* O *such that* $\langle \mathbf{O} \rangle_{\rho(M,g)} = \langle \mathbf{O} \rangle_{\rho(M',g')}$ *for two distinct spacetimes* (M, g) *and* (M', g')*, with* $\langle . \rangle_{\rho(M,g)}$ *taken over* $\mathfrak{C}_\rho(M, g)$*, then these spacetimes are indistinguishable with respect to a spacetime covariant observable* $\mathcal{G}$ *where* $O \sim_\rho \mathcal{G}$*, at scales* $\gg \rho^{-1}$*.*

Thus, the longer the list of geometric order invariants corresponding to manifold invariants, the closer we are to proving the full Hauptvermutung.

2.4 Summary

In this chapter we introduced the main CST hypothesis where the space of continuum spacetimes is replaced by a set of discrete geometries which are locally finite posets. We discussed several aspects of the continuum approximation which is inspired by the HKMMKP theorems. An important result is that the random discreteness, which characterises continuumlike causal sets, does not violate Lorentz invariance when $C \sim_\rho \mathbb{M}^d$. As a result, causal sets possess a characteristic nonlocality which distinguish CST from other discrete approaches to quantum gravity.

That there is a fundamental discrete randomness even *kinematically* is not always easy for a newcomer to CST to come to terms with. Even before imposing a quantum dynamics, the continuum approximation of CST is probabilistic, as a kinematic, stochastic, discrete geometry. Whatever the quantum theory of causal sets, any statement about the emergence of the continuum must be compatible with this continuum approximation. To say that a quantum observable "looks" like a continuum geometric or topological invariant, it must strictly satisfy the discrete-continuum correspondence of CST.

We emphasise once again that in CST it is not the discrete substructure, i.e., the causal set which is considered to be an approximation, but the converse. The continuum is considered to be an emergent macroscopic entity arising from an underlying ensemble of causal sets $\mathfrak{C}_\rho(M, g)$. It is in this sense that the continuum approximation can be viewed as a first quantisation, before the full quantum dynamics is implemented. The fluid analogy is handy. Even without a dynamics, underlying the macroscopic fluid is an ensemble of molecular-level configurations corresponding to the same kinematic macroscopic state. An ensemble of causal sets $\mathfrak{C}_\rho(M, g)$ is similarly approximated by the spacetime (M, g). And, just as the set of allowed microstates of a fluid depends (among other things) on the molecular scale, the causal set "microstates" (i.e., the ensemble $\mathfrak{C}_\rho(M, g)$) of a given continuum spacetime (M, g) depend on the **discreteness scale** ρ^{-1}, which is a fundamental spacetime volume cut-off. The most obvious choice for ρ^{-1} is the Planck spacetime volume, but depending on the situation, we may want to impose this cut-off at larger scales as well.

The sequence of arguments that lead up to CST is tight: from being inspired by the HKMMKP theorem, which gives us conformal geometry, we arrive at its discrete version, which gives us the whole spacetime, albeit only approximately. In the process, the discrete number-volume correspondence, leads us to the Poisson generated ensemble of causal sets, which in turn implies to Lorentz invariance! The phenomenological consequences of Lorentz invariance and the cosmological constant prediction which we will recall in Chap. 6, the choice of the Poisson sprinkling for implementing the continuum approximation appears to be "spot on"!

Kinematics

3

Abstract

Causal set theory posits that any continuum causal spacetime (M, g) is an approximation to an underlying discrete structure, namely, an ensemble $\mathfrak{C}_\rho(M, g)$ of causal sets generated via a generalised Poisson point process or "Poisson sprinkling" into (M, g) at fixed density ρ. If this hypothesis is correct, $\mathfrak{C}_\rho(M, g)$ should contain all the topological and geometric information in (M, g) at scales larger than the discreteness scale ρ^{-1}. In this chapter we show several examples of the "reconstruction" of continuum geometric and topological quantities from the underlying causal set ensemble. These are constructed from geometric order invariants which are the discrete covariant observables of the theory.

3.1 Introduction

Any viable theory of quantum gravity should be able to reproduce known physics at the relevant scales. At the level of kinematics, this means that physically important classical spacetimes must be included in the space of all quantum geometries. These include amongst others, Minkowski, de Sitter, FRW cosmologies and observationally relevant blackhole spacetimes. In a fundamentally discrete theory like CST, this means that the continuum approximation of these spacetimes, or equivalently, the underlying causal set ensemble, must be included in the theory. It is therefore important to explicitly show that the $\mathfrak{C}_\rho(M, g)$ in fact contains all geometrical and topological information about (M, g) at scales sufficiently larger than the discreteness scale ρ^{-1}. In this chapter we present several geometric reconstruction results where order invariants in the ensemble correspond, on average, to various geometric and topological invariants of the continuum, upto the discreteness scale. As discussed in Chap. 2, the fundamental conjecture or Hauptvermutung of CST, requires that $\mathfrak{C}_\rho(M, g)$ reproduces (M, g) "approximately uniquely", i.e., "upto scale ρ^{-1}". The reconstruction results thus prove weaker, observable dependent versions of this conjecture.

S. Surya, *The Causal Set Approach to Quantum Gravity*, Lecture Notes in Physics 1036,
https://doi.org/10.1007/978-3-031-84420-1_3

A useful analogy to keep in mind as we traverse this chapter, is that of a box of ideal gas characterised by its macroscopic observables – volume, pressure, temperature and number of particles. These are determined by the underlying microscopic ensemble of different configurations of the gas. This ensemble therefore cannot be associated with vastly different values of the macroscopic observables. Hence, upto the characteristic scale of the system, the macroscopic observables are entirely, and uniquely, captured by the microscopic ensemble of the gas molecules. In CST, similarly, the "macroscopic" geometric and topological observables of a spacetime (M, g) – for example, the spacetime dimension or the spatial homology – are conjectured to be entirely captured by the underlying ensemble of causal sets $\mathfrak{C}_\rho(M, g)$, at scales much larger than the discreteness scale. Thus, only one set of these macroscopic spacetime observables should be associated with every such ensemble.

Establishing an explicit correspondence between an order invariant O and a macroscopic spacetime invariant G at a scale ρ^{-1} implies that the expectation value $\langle \mathbf{O} \rangle$ over $\mathfrak{C}_\rho(M, g)$ cannot be obtained from spacetimes with a significantly different macroscopic invariant G'. This gives us several weak forms of the Hauptvermutung for each $O \sim_\rho G$ correspondence. In the absence of a "complete" list of spacetime observables, this provides supporting evidence for, and brings us closer, at least in spirit, to the Hauptvermutung.

Since order invariance is a discrete version of covariance, a reconstruction of the continuum geometry using order invariants ensures that the resulting continuum entity is covariant. More generally, order invariants are the CST analogues of the covariant observables of quantum gravity. The class of "geometric" order invariants, i.e., those for which $O \sim_\rho G$, provide an important class of discrete observables, essential to both the construction of a quantum dynamics, as well as the interpretion of the results, as we will see in Chap. 5.

The main challenge in reconstructing the spacetime (M, g) from $\mathfrak{C}_\rho(M, g)$ lies in the correct identification of microscopic order invariants that correspond to specific macroscopic spacetime observables. Because of the nonlocality inherent in continuumlike causal sets, standard lattice and simplicial geometry methods are not useful and there is a need for novel order theoretic constructions. These constructions in turn challenge our Riemannian "intuition" of distance, and bring in a fresh perspective on Lorentzian geometry, based entirely on causality and volume.[1]

This is a fun chapter to "wet ones feet" with causal sets, since at heart it is purely geometric and hence conceptually the simplest. The reconstructions force us to rethink our Riemannian-schooled brains and rewire them to the Lorentzian framework. A challenge for the interested reader is to explore alternative order invariants for the cases we present in this chapter.

After a broad discussion of the reconstruction of spacetime geometry and topology in Sect. 3.2, we focus our attention on three classes of order invariants obtained from

[1] Riemannian "intuition", it can be argued, stems from a deeply rooted elementary school training in Euclidean geometry, and has less to do with an innate sense of space over spacetime, but more with mathematical conditioning.

chains (Sect. 3.3), antichains (Sect. 3.4) and order intervals (Sect. 3.5.) We show how these simple order invariants, in various combinations, give rise to geometric order invariants and hence contain a rich amount of geometric and topological information about the continuum. The aim in presenting these specific order invariants is for the sake of pedagogical simplicity, but it is by no means an exhaustive list.

3.2 Geometric Reconstruction

We begin by setting up the framework, as well as the notation and terminology that we will require in the following Sections.

A "complete" set of diffeomorphism invariant observables is notoriously difficult to find in the continuum. However, an important class of such observables are (i) topological invariants like the spacetime dimension and (ii) integrated geometric quantities like the action (for finite volume spacetimes.) We will refer to such covariantly defined observables collectively as **spacetime observables**.

We wish to associate with such a spacetime observable G, a corresponding random variable $\mathbf{O}$ in $\mathfrak{C}_\rho(M,g)$ where O is an order invariant. The correspondence is generically dependent on ρ, and we denote this as $O \sim_\rho G$. As defined in Chap. 2, an order invariant can be defined over any causal set, not just one that comes from a sprinkling. Most order invariants do not have a geometric interpretation. The aim of geometric reconstruction is to identify those that do, by using explicit features of $\mathfrak{C}_\rho(M,g)$.

In the following sections we will construct order invariants whose expectation values can be identified with spacetime observables,

$$\langle \mathbf{O} \rangle \sim_\rho G. \tag{3.1}$$

Because of the nonlocality of continuumlike causal sets, there is no local tangent space and therefore standard discretisation tools are not helpful. The identification of geometric order invariants is therefore often non-trivial, and in practice requires a long series of guesses and verifications. In turn this has resulted in new causal geometric "tools" which are useful for analysing aspects of continuum causal spacetime geometry.

In certain cases, an analytic calculation yields an exact result in the continuum limit, i.e.,

$$\lim_{\rho\to\infty} \langle \mathbf{O} \rangle \sim_\rho G, \tag{3.2}$$

while in others, the identification is checked numerically. In general, numerical simulations play an important role in geometric reconstruction not only in honing one's initial intuition, but also in supplementing analytic results. Corrections to the expectation value for large but finite ρ are physically important, but are not always amenable to analytical calculations. Of interest are fluctuations about the mean which quantify how a single realisation in $\mathfrak{C}_\rho(M,g)$ fares.

Additionally in some cases, the continuum correspondence for a given geometric order invariant is known only for a specific class of spacetimes. The simplest cases to prove analytically are for causal sets approximated by $\mathbb{M}^d$. In some cases, these have been generalised to Riemann normal neighbourhoods (RNN) of generic spacetimes, and in other cases, to either conformally flat or constant curvature spacetimes, like de Sitter and anti de Sitter spacetimes.

In addition to such spacetime observables, we are also interested in covariantly defined geometric properties associated with a set of fixed events, for example, the proper time between pairs of events and the scalar curvature at a single event. While these may not always correspond to global geometric observables, they give us insights into how order and number encode local geometric properties.

For the sake of pedagogy, our focus in this chapter will be on building order invariants from three types of simple order theoretic structures—chains, antichains and intervals. The hope is to give the reader a working knowledge of geometric reconstruction. For continuumlike causal sets, more than one geometric order invariant could correspond to the same spacetime invariant. For example, there are several flat spacetime dimension estimators, some of which we will encounter in this chapter.

Broadly, the constructions follow the logic:

1. Start with the causal set ensemble $\mathfrak{C}_\rho(M, g)$ associated with the Poisson sprinkling into a spacetime or spacetime region (M, g).
2. Identify within $\mathfrak{C}_\rho(M, g)$ the order invariant O which corresponds to a spacetime observable G.
3. Show that

$$\lim_{\rho\to\infty} c\langle \mathbf{O}\rangle = G \tag{3.3}$$

where $c = \rho^r$ gives the physical dimension of G, assuming O is defined to be dimension free.

Myrheim's work [18], which predates [2], has the earliest examples of such spacetime-order invariants. We will not however follow the chronology of discovery, but rather focus on a set of simple order invariants constructed from chains, antichains and order intervals. A range of spacetime invariants can be constructed from these, thus demonstrating the richness of geometric information encoded in $\mathfrak{C}_\rho(M, g)$.

We now define some more terminology that will be needed as we progress.

A discrete version of a causal curve in C is a totally ordered subset of C, or **chain** (Sect. 2.3.1.) A $k-$**chain** c_k is a totally ordered set of elements which admits a labelling $\{e_1 \ldots\ldots e_k\}$ such that $\forall\, i = \{1, \ldots k-1\}$, $e_i \prec e_{i+1}$. A $k-$chain $\mathcal{P}_k$ is a $k-$**linked-path** if additionally, $\forall\, i = \{1, \ldots k-1\}$, $e_i \prec\!\!* \, e_{i+1}$. We define the **length** of the chain c_k to be k, i.e., the number of elements in c_k. A chain c_k is said to be a **maximal chain** if it is the longest possible chain in C. There need not be a

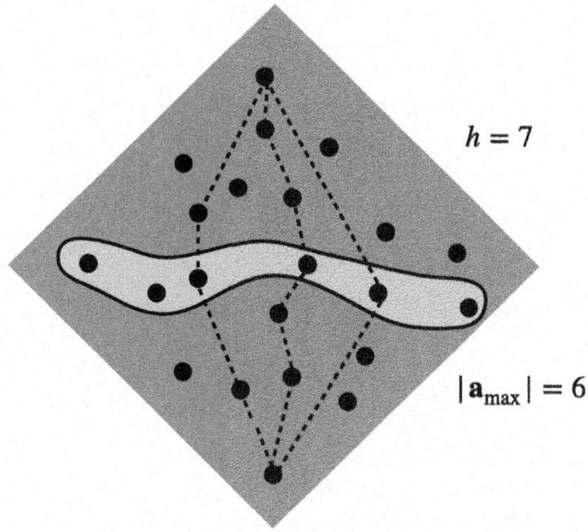

Fig. 3.1 Examples of **chains** and **antichains** in a causal set obtained from a Poisson sprinkling into a causal diamond in $\mathbb{M}^2$. The **height** of the causal set is $h = 7$, which is the length of the longest chain. The **width** of the causal set is $|\mathbf{a}_{max}| = 6$, which is the size of the largest antichain

unique maximal chain.[2] The length of the maximal chain is also called the **height** of C (See Fig. 3.1.)

It is clear that causal sets favour causal relations over spatial ones: to define a spatial curve is far more complicated, since the spacelike relation is characterised by the absence of an order relation. Unlike the chain, there is something already "co-dimension one" about an **antichain** as shown in Fig. 3.1. Hence rather than a "length" we find it more natural to characterise it by its "spatial volume" or cardinality. Instead of the word "maximal" which we prefer to use for being future-most (Eq. (2.19)), we use the term **inextendible antichain** to denote an antichain for which there are no other elements in the causal set which are spacelike to all its elements.

Exercise 3.1 Numerically simulate an ensemble of causal sets using the binomial point process into the causal diamond $\mathbb{D}^2 \subset \mathbb{M}^2$ at a fixed density ρ and fixed total continuum volume V, but without fixing the total number of elements n. A reasonable range of densities to work with is about $10 - 25$ elements per unit 2-volume(area). Find $\langle \mathbf{n} \rangle$ and the fluctuations $\Delta \mathbf{n}$ for an ensemble generated from the sprinklings. Show convergence as the number of trials increases.

We will also use the following notations for continuum quantities. A causal diamond $\mathbb{D}^d$ or Alexandrov interval $I(p,q)$ in $\mathbb{M}^d$ is characterised by its volume $\mathrm{vol}(p,q)$ or equivalently its proper time $\tau(p,q)$. These are related to each other by

$$\mathrm{vol}(p,q) = \zeta_d \tau(p,q)^d \tag{3.4}$$

[2] There is room for confusion here. We use the word maximal here not as future-most, as in Eq. (2.19), but as being the longest. We will refrain from this confusing terminology as far as possible in the rest of this book and use it only to mean future-most.

where

$$\zeta_d \equiv S_{d-2}\frac{2^{1-d}}{d(d-1)} \tag{3.5}$$

with

$$S_{d-2} \equiv \frac{2\pi^{\frac{d-1}{2}}}{\Gamma\left(\frac{d-1}{2}\right)} \tag{3.6}$$

being the volume of the unit $(d-2)$-dimensional sphere $\mathbb{S}^{d-2}$.

3.3 Chains

The order relation $e_1 \prec e_2$ between elements is most basic to a causal set. The number R of such causal **relations** in a finite causal set is the simplest order invariant we can construct. Its simplicity is however deceptive, since it encodes a very important dimension estimator, the Myrheim-Meyer flat spacetime dimension estimator for $\mathfrak{C}_\rho(\mathbb{M}^d)$.

The **ordering fraction** is the ratio of the number of relations R to the maximum number, $\binom{n}{2}$, of possible relations in an n-element causal set,

$$\mathbf{r} \equiv \frac{R}{\binom{n}{2}}. \tag{3.7}$$

Exercise 3.2 Show that the ordering fraction takes on its maximum value only for a totally ordered set or chain on n-elements.

Exercise 3.3 Show that for the set of $n = 3$ causal sets, R serves to distinguish all but the time-symmetric causal sets. Use the causal sets in Appendix A as a guide.

As we have seen, the binary relation is easily extended to the triplet relation or 3-chain $e_1 \prec e_2 \prec e_3$, and more generally to a k-chain $\mathrm{c}_k, e_1 \prec e_2 \prec \ldots \prec e_k$, where by transitivity, $e_i \prec e_j,\ \forall j \geq i+1$. This is a totally ordered subset of C, and every k-chain $\mathrm{c}_k \subset C$ contains $\binom{k}{l}$ number of l-element sub-chains (see Fig. 3.2.) While each labelled chain is not an order invariant, the number of k-chains in C is an order invariant, as is its height h. In a continuumlike causal set, for $d \geq 2$, like that shown in Fig. 2.9, C contains elements that are spacelike related to each other and hence $h < n$.

In $\mathbb{M}^d$, the timelike distance $\tau(p, q)$ between two chronologically related events $p \ll q$ is obtained by *maximising* the distance over timelike paths from p to q. Thus we expect that $\langle \mathbf{h} \rangle$ for $\mathfrak{C}_\rho(\mathbb{M}^d)$ is proportional to the proper time between these events, with the constant of proportionality dependent on $\rho^{\frac{1}{d}}$, as suggested by Myrheim [18].

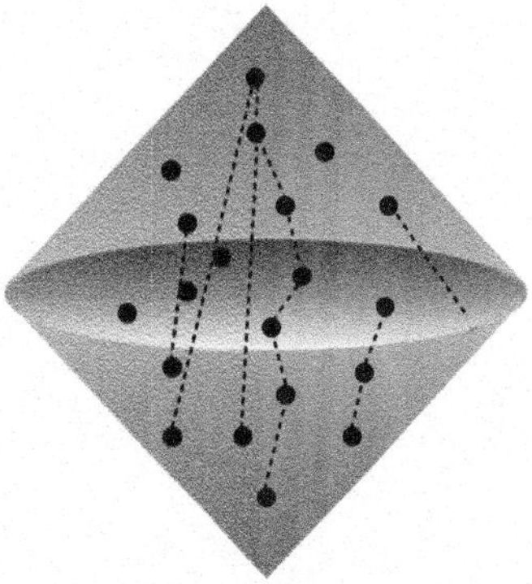

Fig. 3.2 Chains of different length in $\mathbb{M}^3$

Using results in order theory [110] Brightwell and Graham showed in [35] that for a causal diamond $\mathbb{D}^d$ of volume V

$$\lim_{\rho\to\infty} \frac{\langle \mathbf{h} \rangle}{(\rho V)^{1/d}} = m_d \tag{3.8}$$

where m_d is a dimension dependent constant with $m_2 = 2$ and

$$1.77 \le \frac{2^{1-\frac{1}{d}}}{\Gamma(1+\frac{1}{d})} \le m_d \le \frac{2^{1-\frac{1}{d}} e\,(\Gamma(1+d))^{\frac{1}{d}}}{d} \le 2.62, \quad d \ge 3. \tag{3.9}$$

Since the proper time between the end points of $\mathbb{D}^d$, is $\tau = (V/\zeta_d)^{1/d}$ (Eq. (3.4)) this gives us a correspondence between $\langle \mathbf{h} \rangle$ and τ. For finite ρ, however, there are large fluctuations in $\mathbf{h}$ and hence the correspondence becomes meaningful only when averaged over a large ensemble [111].[3]

We will consider $\mathfrak{C}_\rho(M, g)$ for spacetimes (M, g) which are themselves Alexandrov intervals in $\mathbb{M}^d$ or more generally, Riemann Normal Neighbourhoods (RNN) in any curved spacetime. As we will see, the number C_k of k-chains c_k in C is a family of order invariants which proves to be a treasure-house of information about the associated spacetime geometry.

We begin with $\mathfrak{C}_\rho(\mathbb{D}^d)$ where $\mathbb{D}^d$ is a causal diamond or Alexandrov interval in $\mathbb{M}^d$. The expectation value for the number of elements or equivalently the number of 1-chains C_1 in $\mathbb{D}^d$ is

$$\langle \mathbf{C}_1 \rangle = \rho \int_{\mathbb{D}^d} dV_x = \rho \mathrm{vol}(\mathbb{D}^d). \tag{3.10}$$

[3] An interesting connection exists between the length of the longest chain and the flight boarding problem [112].

In order to calculate $\langle \mathbf{C}_2 \rangle$, we note that once the first element has been fixed at some $x_1 \in \mathbb{D}^d$, any $x_2 \in J^+(x) \cap \mathbb{D}^d$ satisfies $x_1 \prec x_2$. Thus

$$\langle \mathbf{C}_2 \rangle = \rho^2 \int_{\mathbb{D}^d} dV_{x_1} \int_{J^+(x) \cap \mathbb{D}^d} dV_{x_2}. \tag{3.11}$$

Generalising, the average number of k-chains $\mathbf{c}_k$ is given by the nested sequence of integrals

$$\langle \mathbf{C}_k \rangle = \rho^k \int_{\mathbb{D}^d} dV_{x_1} \int_{J^+(x_1) \cap \mathbb{D}^d} dV_{x_2} \ldots \int_{J^+(x_{k-1}) \cap \mathbb{D}^d} dV_{x_k} = \rho \int_{\mathbb{D}^d} dV_{x_1} \langle \mathbf{C}_{k-1} \rangle, \tag{3.12}$$

over a nested sequence of intervals

$$J^+(x_{k-1}) \cap \mathbb{D}^d \subset J^+(x_{k-2}) \cap \mathbb{D}^d \ldots \subset J^+(x_1) \cap \mathbb{D}^d \subset \mathbb{D}^d. \tag{3.13}$$

The right-most integral can be evaluated using the Lorentz invariance of the spacetime volume, namely by choosing "upright" conical coordinates (t, r, Ω_{d-2}) where Ω_{d-2} denotes the $d-2$ angular coordinates on the conical cross-section. Choosing coordinates so that $\mathbb{D}^d = I(p, q)$ with $p = (-T/2, 0, \ldots 0)$, $q = (T/2, 0, \ldots 0)$,

$$\langle \mathbf{C}_1 \rangle = 2\rho \int_0^{T/2} dt \int_0^{T/2-t} dr r^{d-2} \int_{\mathbb{S}^{d-2}} d\Omega_{d-2} = \rho \zeta_d T^d \equiv \rho \mathbf{V}_0, \tag{3.14}$$

where ζ_d is given in Eq. (3.5). Nesting this in the next integral and using light cone coordinates (u, v, Ω_{d-2})

$$\begin{aligned} \langle \mathbf{C}_2 \rangle &= \rho^2 \zeta_d \int_{\mathbb{D}^d} dV_{x_1} T_1^d \\ &= \rho^2 \zeta_d \int_{-\frac{T}{2}}^{\frac{T}{2}} dv_1 \int_{-\frac{T}{2}}^{v_1} du_1 \int d\Omega_{d-2} \sqrt{-\eta(u, v)} T_1^m, \end{aligned} \tag{3.15}$$

where T_1 is the proper time from x_1 to q and

$$\sqrt{-\eta(u, v)} = \frac{(u-v)^{d-2}}{2^{d-1}} \left(\prod_{i=1}^{d-3} \sin^{d-i-2} \theta_i \right). \tag{3.16}$$

This can be evaluated by recasting $u - v = (\frac{T}{2} - v) - (\frac{T}{2} - u)$ and using the binomial expansion to get

$$\langle \mathbf{C}_2 \rangle = \rho^2 \mathbf{V}_0^2 \frac{\Gamma(d+1)\Gamma(d/2)}{4\Gamma(3d/2)}. \tag{3.17}$$

Using Eq. (3.12) we can evaluate $\langle \mathbf{C}_k \rangle$ inductively to get

$$\langle \mathbf{C}_k \rangle = \frac{\rho^k \zeta_d^k}{k} \left(\frac{\Gamma(d+1)}{2} \right)^{k-1} \frac{\Gamma(\frac{d}{2})\Gamma(d)}{\Gamma(\frac{kd}{2})\Gamma(\frac{(k+1)d}{2})} T^{kd} \tag{3.18}$$

Exercise 3.4 Complete the induction argument to show Eq. (3.18).

Noticing that the dependence on the dimension d factors out, we thus find a family of **dimension estimators**

$$f_{(k_1,k_2)}(d) \equiv \frac{\langle \mathbf{C}_{k_1}\rangle^{k_2}}{\langle \mathbf{C}_{k_2}\rangle^{k_1}} = \left(\frac{\Gamma(d+1)}{2\Gamma(d)\Gamma(\frac{d}{2})}\right)^{k_1-k_2} \frac{\left(k_2\Gamma(\frac{k_2 d}{2})\Gamma(\frac{(k_2+1)d}{2})\right)^{k_1}}{\left(k_1\Gamma(\frac{k_1 d}{2})\Gamma(\frac{(k_1+1)d}{2})\right)^{k_2}}, \tag{3.19}$$

which, for every pair (k_1, k_2), are functions only of the dimension [33]. The special case

$$f_{(2,1)}(d) = \frac{\Gamma(d+1)\Gamma(d/2)}{4\Gamma(3d/2)} \tag{3.20}$$

is referred to as the flat spacetime **Myrheim-Meyer dimension estimator**. Figure 3.3 shows $f_{(2,1)}(d)$ as a function of d. For large n, it is half the ordering fraction. In the special case $d = 2$, $f_{2,1}(d) = 1/4$, so the ordering fraction is $\frac{1}{2}$. Numerical simulations of this dimension estimator in different dimensions can be found in [113].

Equivalently, for a fixed ρ, the distribution of C_k with k is a characteristic function of the dimension, as demonstrated by numerical simulations in [114].

Exercise 3.5 We define a Kleitmann-Rothschild (KR) type poset C_{KR} to be: (i) a three layer poset $C_{KR} = L_1 \sqcup L_2 \sqcup L_3$, where each L_i is an antichain, (ii) such that there are exactly $n/4$ elements each in L_1 and L_3 and exactly $n/2$ elements in L_2, (iii) every element in L_1 is to the past of every element in L_3 (iv) every element in L_1 is linked to exactly $n/2$ elements in L_2 and (v) every element in L_3 is linked to exactly $n/2$ elements in L_2. Using the dimension estimator $f_{2,1}$ show that C_{KR} has a non-integer dimension.

These calculations can be generalised to an Alexandrov interval $I(p, q)$ contained in an RNN of a generic spacetime. In [115] it was shown that

$$\langle \mathbf{C}_k\rangle = \langle \mathbf{C}_k^{(\eta)}\rangle \left[1 + (\alpha_k(d)R(0) + \beta_k(d)R_{00}(0))T^2\right] + O(T^{kd+3}), \tag{3.21}$$

where $\langle \mathbf{C}_k^{(\eta)}\rangle$ is the expectation value of k-chains in flat spacetime (Eq. (3.18)) and

$$\alpha_k(d) = -\frac{dk}{12(kd+2)((k+1)d+2)}$$
$$\beta_k(d) = \frac{dk}{12((k+1)d+2)}. \tag{3.22}$$

What is immediately obvious from this expression is that one can extract *all* the spacetime observables that appear in Eq. 3.21 by using different k values. Thus, in

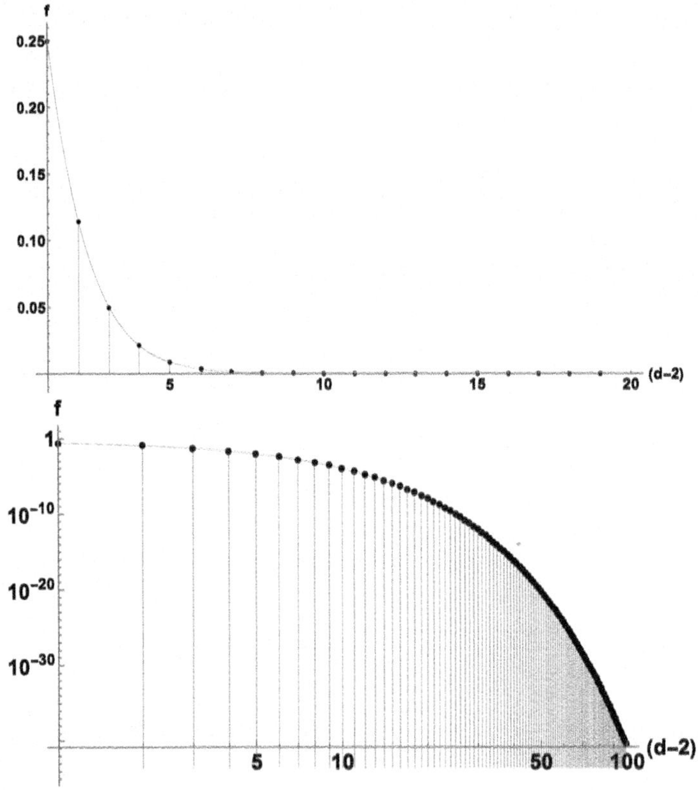

Fig. 3.3 Myrheim-Meyer dimension estimator $f_{(2,1)}(d)$ versus d. The second figure on the right is a Log-Log plot

addition to the dimension d, $\langle \mathbf{C}_k \rangle$ gives us the scalar curvature $R(0)$, the Ricci tensor component $R_{00}(0)$, as well as the proper time T of the interval. Since these appear in the expression for $\langle \mathbf{C}_k \rangle$ for *all* k, they can be all be obtained using any four distinct values of k.

To begin with, let us fix the dimension d to find an alternative expression for the proper time of the interval

$$T^{3d} = \frac{1}{2d^2}\Big(J_1 - 2J_2 + J_3\Big). \tag{3.23}$$

where

$$J_k \equiv (kd+2)((k+1)d+2))\frac{1}{\zeta_d^3}\left(\frac{\langle \mathbf{C}_k \rangle}{\rho^k \chi_k}\right)^{3/k}, \quad \chi_k \equiv \frac{1}{k}\left(\frac{\Gamma(d+1)}{2}\right)^{k-1} \frac{\Gamma(\frac{d}{2})\Gamma(d)}{\Gamma(\frac{kd}{2})\Gamma(\frac{(k+1)d}{2})}. \tag{3.24}$$

This is only one of a family of proper time estimators that can be constructed starting with three different k values. Unlike the Brightwell-Gregory definition, this estimator cannot be meaningfully defined on a single causal set, since one must first compute the averages of the C_k and then take their $3/k$th power. This is of course also the case with the dimension estimator Eq. (3.19). However, it is possible to circumvent this by substituting ρV with n (instead of on average). Simulations suggest that the fluctuations are suppressed in comparison to the Brightwell-Gregory proposal, but this has yet to be firmly established.

In addition, we obtain expressions for the curvatures at the center of $I(p, q)$

$$R(0) = -2(n+2)(2n+2)(3n+2)2^{\frac{3n+2}{3n}} n^{\frac{4}{3n}-1} \frac{(K_1 - 2K_2 + K_3)}{(J_1 - 2J_2 + J_3)^{\frac{3n+2}{3n}}}$$
$$R_{00}(0) = -\frac{4(2n+2)(3n+2)}{2^{\frac{3n+2}{3n}} n^{\frac{4}{3n}-1}} \frac{((n+2)Q_1 - (5n+4)Q_2 + (4n+2)Q_3)}{(J_1 - 2J_2 + J_3)^{\frac{3n+2}{3n}}}. \tag{3.25}$$

where

$$Q_k \equiv \frac{1}{\zeta_d^3} \left(\frac{\langle \mathbf{C}_k \rangle}{\rho^k \chi_k} \right)^{3/k}$$
$$K_k \equiv ((k+1)n + 2) Q_k \tag{3.26}$$

which allows us to re-express $J_k = (kn+2)K_k$.

Again, we can construct a family of estimators starting with any three values of k for a fixed dimension. In order to find an RNN dimension estimator from C_k, we need four values of k. Solving for d we find the implicit functional form for the first four values of k.

$$(d+2)(2d+2)\left(\frac{\langle \mathbf{C}_1 \rangle}{\chi_1}\right)^4 - 3(2d+2)(3d+2)\left(\frac{\langle \mathbf{C}_2 \rangle}{\chi_2}\right)^2$$
$$+3(3d+2)(4d+2)\left(\frac{\langle \mathbf{C}_3 \rangle}{\chi_3}\right)^{4/3} - (4d+2)(5d+2)\left(\frac{\langle \mathbf{C}_4 \rangle}{\chi_4}\right) = 0 \tag{3.27}$$

The results of recent simulations show that these curvature estimates reproduce the continuum values with high accuracy in $d = 2$ de Sitter spacetime and $d = 3$ FLRW universes with flat spatial sections [114].

In [116] a **chain action** was defined by identifying n with $\rho \zeta_d T^{dk}$ (rather than on average.) This identification allows us to rewrite

$$C_k^{(\eta)} \approx \frac{n^k}{k} \left(\frac{\Gamma(d+1)}{2} \right)^{k-1} \frac{\Gamma(\frac{d}{2})\Gamma(d)}{\Gamma(\frac{kd}{2})\Gamma(\frac{(k+1)d}{2})} = n^k A(k, d)$$
$$C_k \approx n^k A(k, d) \left[1 + \left(\frac{n}{\rho \zeta_d} \right)^{2/d} (\alpha_k(d) R(0) + \beta_k(d) R_{00}(0)) \right] + O(n^{k+3/d}), \tag{3.28}$$

from which, for any choice of k_1, k_2

$$R = \left(\frac{\rho}{n}\right)^{2/d}\left(B(k_1, k_2, d)\frac{C_{k_1}}{n^{k_1}} + B(k_2, k_1, d)\frac{C_{k_2}}{n^{k_2}} + C(k_1, k_2, d)\right), \tag{3.29}$$

where

$$\begin{aligned} B(k_1, k_2, d) &\equiv \zeta_d^{2/d}\frac{\beta_{k_2}(d)}{A(k_1, d)} \times \frac{1}{(\beta_{k_2}(d)\alpha_{k_1}(d) - \beta_{k_1}(d)\alpha_{k_2}(d))}, \\ C(k_1, k_2, d) &\equiv \zeta_d^{2/d}\frac{(\beta_{k_1}(d) - \beta_{k_2}(d))}{(\beta_{k_2}(d)\alpha_{k_1}(d) - \beta_{k_1}(d)\alpha_{k_2}(d))} \end{aligned} \tag{3.30}$$

Given that R is constant over the RNN, this gives the following form for the discrete action:

$$S_C \equiv R \times \frac{n}{\rho} = \equiv \left(\frac{\rho}{n}\right)^{2/d-1}\left(B(k_1, k_2, d)\frac{C_{k_1}}{n^{k_1}} + B(k_2, k_1, d)\frac{C_{k_2}}{n^{k_2} + C(k_1, k_2, d)}\right). \tag{3.31}$$

In Sect. 3.5 we will find a very different kind of action, the BDG action using order intervals. For that action, one can argue that in continuumlike causal sets the "far away" contributions to the curvature at a point cancel out and only quasilocal contributions matter. Whether this is possible for S_C is an open question, and is well worth studying. In [116] a modified chain action was studied in $\mathbb{M}^2$ and in 2d de Sitter and anti de Sitter spacetimes, and shown to be "as good as" the BDG action, but with smaller fluctuations, which is promising.

What else can we learn from chains? As we will show in the next chapter, the $d = 2$ quantum free scalar field theory on $C \sim_\rho \mathbb{D}^2$ can be entirely determined from C_k! This is particularly simple to see for the massless free scalar field theory, where the retarded Green function takes the form

$$G_0^{(2)}(p, q) = \frac{1}{2}\theta(t_p - t_q)\theta(\tau^2(p, q)). \tag{3.32}$$

For a causal set in $\mathfrak{C}_\rho(\mathbb{D}^2)$, therefore we expect that $G_0^{(2)}(e_1, e_2) = \frac{1}{2}$ if $e_1 \prec e_2$ and zero otherwise, i.e.,

$$G_0^{(2)}(x, x') = \frac{1}{2}M_C(x, x'), \tag{3.33}$$

where the **causal matrix**

$$M_C(e, e') \equiv \begin{cases} 1 \text{ if } e' \prec e \\ 0 \text{ otherwise} \end{cases} \tag{3.34}$$

Using a formal expansion of the massive Green function in terms of the massless one (see Chap. 4) it can be shown that the massive scalar Green function in $d = 2$ takes the form

$$K_m^{(2)}(e_1, e_2) = \sum_{k=0}^{\infty} \left(-\frac{m^2}{\rho}\right)^k \left(\frac{1}{2}\right)^{k+1} C_k(e_1, e_2) \tag{3.35}$$

where the sum terminates for each pair e_1 and e_2 in C.

Exercise 3.6 Show that the number of $k \geq 2$-chains from $e_i, e_j \in C \in \mathfrak{C}_\rho(\mathbb{D}^2)$ is given by $(M_C^{k-1})_{ij}$.

As we have seen in this section, the simple order invariants C_k are rich with continuum information. We end this sub-section on chains by speculating on the possibility that they also contain continuum topological information. The set of k-chains in a finite element causal set C can be used to construct an **order complex** as follows. 0-simplices are the elements of C and the $k-1$-simplices are the k-chains. Every k-simplex is therefore made up of $\binom{k}{l}$ number of l-simplices and is homological to the k-ball $\mathcal{B}^k$. For a causal set that embeds into the $d = 2$ flat cylinder spacetime, for example, one would expect there to be "crown" sub-posets which wrap around the spatial $\mathbb{S}^1$ and capture the topological non-triviality of the cylinder. A $2m$ element crown poset $\mathbf{cr}_{2m}$ is a **bilayer** poset which is a disjoint union of two m-antichains $L_1 = \mathrm{a}_m, L_2 = \mathrm{a}_m$, such that every element $e_i^{(1)}$ in the bottom antichain (or layer) L_1 is connected to precisely two elements $e_i^{(2)}$ and $e_{i+1}^{(2)}$, in the upper antichain (or layer) L_2, with the identification $i + m \sim i$ as shown in Fig. 3.4. The simplicial complex of $\mathbf{cr}_{2m}$ can be shown to be homological to $\mathbb{S}^1$ and hence to the spacetime (see the following exercise.)

Exercise 3.7 Show that the crown poset $\mathbf{cr}_6$ is homological to $\mathbb{S}^1$ using the chain simplicial complex described above. Show that this is also true for a stacked tower of crowns, i.e., a poset with three layers $\{L_1, L_2, L_3\}$ with $|L_i| = 3$ and such that each of the pairs $\{L_1, L_2\}$ and $\{L_2, L_3\}$ are the layers of two separate crowns $\mathbf{cr}_6$ with the relations between elements in L_1 and L_3 induced by transitivity.

Exercise 3.8 Construct the chain homology for the KR poset in Fig. A.7.

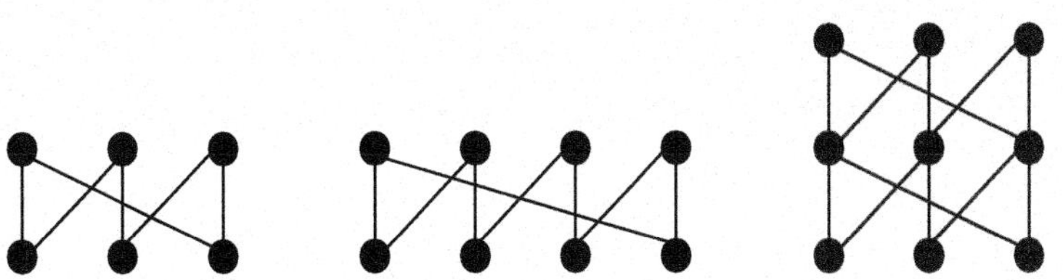

Fig. 3.4 Crown posets $\mathbf{cr}_{2k}$ for $k = 3, 4$ and a small tower of $k = 3$ crowns

3.4 Antichains

Spatial relations in C are defined as the *absence* of causal relations. Thus, in an essential sense, spatial geometry is subsidiary, or derived, from the causal geometry.

The analog of a spatial hypersurface in a causal set C is an antichain $\mathcal{A}$ which is a collection of mutually unrelated elements. For a globally hyperbolic spacetime (M, g) with Cauchy hypersurface Σ, we know that for every $p \notin \Sigma$, either $p \in I^+(\Sigma)$ or $p \in I^-(\Sigma)$. Since this is a causality condition, it can be implemented in the causal set C by requiring $\mathcal{A}$ to be an inextendible antichain, i.e., there is no other element in C which is unrelated to all the elements in $\mathcal{A}$.

Despite this apparent similarity, there is a crucial point of departure from the causal property of a Cauchy hypersurface, due to the nonlocal nature of causal sets. A Cauchy hypersurface in (M, g) is a summary of its past and contains all the information for subsequent evolution of the spacetime. In the continuum, this is captured by the requirement that every future and past endless causal curve intersects Σ precisely once. In a causal set, a future and past endless causal curve is represented by a future and past inextendible chain, for which the relation between every consecutive pair is a link, i.e., a linked-path. Is it the case that every future and past inextendible linked-path[4] in $C \sim_\rho (M, g)$ intersects an inextendible antichain $\mathcal{A}$?

Consider a faithful embedding $\varphi : C \hookrightarrow_\rho (M, g)$ and let $\mathcal{A}$ be an inextendible antichain in C. As we noted in Sect. 2.3.2.1, for a given $p \in \varphi(C)$, the joint probability for a $q \in J^+(p)$ to be in $\varphi(C)$ and to be in the link set $L^+(p)$ is given by $dV_q e^{-\rho \mathrm{vol}(p,q)}$ and hence depends only on $\mathrm{vol}(p, q)$, which is constant along invariant hyperboloids. The probability that an element $p \in \varphi(C) \cap \mathrm{Past}(\varphi(\mathcal{A}))$ is linked to an element $q \in \varphi(C)$ which lies in $L^+(p) \cap \mathrm{Fut}(\varphi(\mathcal{A}))$ is therefore $dV_q e^{-\rho \mathrm{vol}(p,q)}$, where now dV_q lies in $J^+(p) \cap J^+(\varphi(\mathcal{A}))$.

In $\mathbb{M}^d$, for example, such a q will almost surely exist, and hence there is a linked-path that is future and past inextendible containing p, q with $p \prec\!* \, q$ which does not intersect $\mathcal{A}$. It is easy to see that this is true in general for any (M, g) and a generic antichain $\mathcal{A}$ in C. Therefore $\mathcal{A}$ is *not* a summary of its past (see Fig. 3.5)—an inextendible antichain is more like a "sieve" than a Cauchy hypersurface!

Exercise 3.9 Using the causal set C in Fig. 2.9 find an inextendible antichain $\mathcal{A}$ and a pair of linked elements, one to the past of $\mathcal{A}$ and one to its future. Find a chain from a past most to a future most element in C containing this linked pair of elements which therefore never intersects $\mathcal{A}$.

Exercise 3.10 Find the set of all inextendible antichains in the $n = 3$ element crown poset and the $n = 4$ element crown poset. Are these all of the same cardinality?

---

[4] By inextendible we mean that no extension of this path is possible (as a path), to the future or the past, by adding in elements of the causal set. This mimics the standard Lorentzian geometry usage of the term for timelike or causal curves.

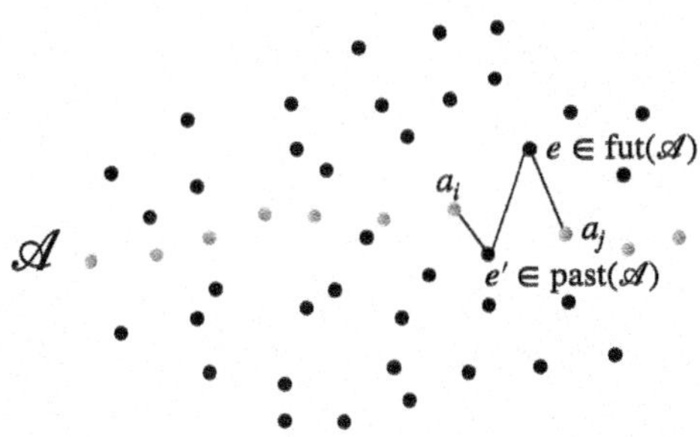

Fig. 3.5 A causal set $C \in \mathfrak{C}_\rho(\mathbb{M}^2)$. An inextendible antichain $\mathcal{A}$ is shown in grey. The elements $e \in \mathrm{Fut}(\mathcal{A})$ and $e' \in \mathrm{Past}(\mathcal{A})$ are such that $e' \prec\!* \, e$. This means that there is no element $a \in \mathcal{A}$ such that $e' \prec a \prec e$, and hence the link from e to e' is "missed" by the inextendible antichain $\mathcal{A}$

Exercise 3.11 A **chain decomposition** $\{c_1, \ldots c_k\}$ of a finite poset C is such that $c_i \subset C$ are chains, and $C = \cup_{i=1}^k c_i$. A minimal chain decomposition is the minimum number of chains required to cover C. Dilworth's theorem [117] states that for any finite poset, the size of the largest (and hence inextendible) antichain in a finite element poset is equal to the size of its minimum chain decomposition. Pick your favorite causal sets from Appendix A and check this.

Despite the non-Cauchy character of $\mathcal{A}$, one can still hope to reconstruct spacelike geometric information from it. $\mathcal{A}$ itself is featureless, since it is a set of disjoint points, but can be given structure by "borrowing" information from its causal neighbourhood. This provides a spatial localisation that is very useful for calculating a broad range of spacetime invariants.

In what follows, we consider a globally hyperbolic spacetime $(M = \Sigma \times \mathbb{R}, g)$ with *compact* Cauchy hypersurface Σ.[5] Let $\varphi : C \hookrightarrow_\rho (M, g)$ be a faithful embedding. Any Cauchy hypersurface Σ divides $\varphi(C)$ into two disjoint parts, $\varphi(C^\pm) \subset J^\pm(\Sigma)$.

Let us henceforth drop the cumbersome notation and identify $\varphi(C^\pm)$ with $C^\pm$, and similarly, $\varphi(\mathcal{A}_\pm)$ with $\mathcal{A}_\pm$. The past-most elements in C^+ and the future-most elements in C^- are inextendible antichains $\mathcal{A}_+$ and $\mathcal{A}_-$ respectively (see Fig. 3.6).

Consider a Gaussian normal neighbourhood (GNN) $\mathcal{N}_\Sigma$ of Σ in which the spacetime metric takes the simplified form

$$ds^2 = -d\tau^2 + h_{ij}(\tau, \mathbf{x})dx^i dx^j, \tag{3.36}$$

where $h_{ij}(\tau, \mathbf{x})$ is the spatial metric on Σ_τ, and τ is the geodesic distance along the unique timelike geodesics from Σ. If $q \in J^+(\Sigma) \cap \mathcal{N}_\Sigma$, then let $\tau^+(\Sigma, q)$ denote the proper time from Σ to q and let $\mathrm{vol}^+(\Sigma, q) \equiv \mathrm{vol}(J^+(\Sigma) \cap J^-(q))$. Similarly, if $q \in J^-(\Sigma) \cap \mathcal{N}_\Sigma$ let $\tau^-(\Sigma, q)$ denote the proper time from q to Σ and let $\mathrm{vol}^-(\Sigma, q) \equiv$

[5] The compactness is important since we need to make probabilistic statements—in a non-compact region, every possibility occurs with certainty!

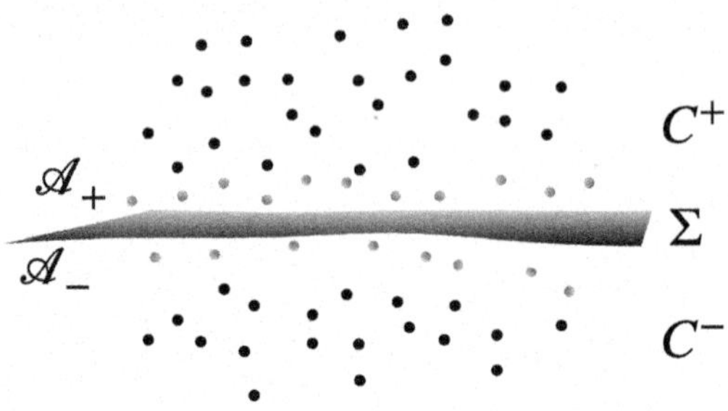

Fig. 3.6 A causal set $C \sim (M, g)$ where (M, g) has compact Cauchy surface Σ, is divided into two disjoint parts $C = C^- \sqcup C^+$, with the former strictly to the past of the latter. $\mathcal{A}_+$ is the minimal (past-most) antichain in C^+ and $\mathcal{A}_-$ is the maximal (future-most) antichain in C^-

$\text{vol}(J^+(q) \cap J^-(\Sigma))$. The continuum "thickness" of $\mathcal{N}_\Sigma$ can then be characterised by one of the two following parameters

$$\begin{aligned} H^\pm(\mathcal{N}_\Sigma) &\equiv \sup_{q \in \mathcal{N}_\Sigma \cap J^\pm(\Sigma)} |\tau^\pm(q, \Sigma)|, \quad H(\mathcal{N}_\Sigma) = \min[H^\pm(\mathcal{N}_\Sigma)], \\ \text{vol}^\pm(\mathcal{N}_\Sigma) &\equiv \sup_{q \in \mathcal{N}_\Sigma \cap J^\pm(\Sigma)} \text{vol}^\pm(\Sigma, q), \quad \text{vol}(\mathcal{N}_\Sigma) = \min[\text{vol}^\pm(\mathcal{N}_\Sigma)]. \end{aligned} \tag{3.37}$$

In addition, consider a Riemann Normal neighbourhood (RNN) $\mathcal{N}(p) \subset \mathcal{N}_\Sigma$ centered at $p \in \Sigma$.

Choose a density ρ such that for $\mathfrak{C}_\rho(M, g)$, $\rho\text{vol}^\pm(\mathcal{N}_\Sigma) >> 1$, which means that $\mathcal{A}_+$ and $\mathcal{A}_-$ lie well within $\mathcal{N}_\Sigma$.

The $\mathcal{A}_\pm$ are devoid of relational information and characterised only by the cardinality $|\mathcal{A}_\pm|$ (which is finite, since Σ is compact). A good starting guess is that $|\mathcal{A}_\pm|$ is a measure of the induced volume of Σ. Defining $P_0 \equiv |\mathcal{A}_+|$, $F_0 \equiv |\mathcal{A}_-|$, the expectation values for the random variables $\mathbf{P}_0$ and $\mathbf{F}_0$ are therefore [118, 119]

$$\begin{aligned} \langle \mathbf{P}_0 \rangle &= \rho \int_{J^+(\Sigma)} dV_p \, e^{-\rho V_\blacktriangle(p)}, \\ \langle \mathbf{F}_0 \rangle &= \rho \int_{J^-(\Sigma)} dV_p \, e^{-\rho V_\blacktriangledown(p)}, \end{aligned} \tag{3.38}$$

where we denote $V_\blacktriangle(p) \equiv \text{vol}(J^+(\Sigma) \cap J^-(p))$ and $V_\blacktriangledown(p) \equiv \text{vol}(J^+(p) \cap J^-(\Sigma))$.

Let us begin with the simplest case of a flat slice $\Sigma \sim I^{d-1}$ in $\mathbb{M}^d$ with zero intrinsic and extrinsic curvature. Using the volume formula Eq. (3.4), the integrals simplify to

$$\begin{aligned} \left\langle \mathbf{P}_0^{\text{flat}} \right\rangle &= \rho \int_\Sigma d^{d-1}x \sqrt{h(0, \mathbf{x})} \int_0^\infty d\tau \, e^{-\rho s_d \tau^d} = \rho \text{vol}(\Sigma) \int_0^\infty d\tau \, e^{-\rho s_d \tau^d}, \\ \left\langle \mathbf{F}_0^{\text{flat}} \right\rangle &= \rho \int_\Sigma d^{d-1}x \sqrt{h(0, \mathbf{x})} \int_{-\infty}^0 d\tau \, e^{-\rho s_d \tau^d} = \rho \text{vol}(\Sigma) \int_{-\infty}^0 d\tau \, e^{-\rho s_d \tau^d}, \end{aligned} \tag{3.39}$$

where $\mathbf{s}_d = 2^{d-1}\zeta_d$ and vol(Σ) denotes the spatial volume of Σ. Hence, just as we had surmised,

$$\mathrm{vol}(\Sigma) = \frac{(\rho \mathbf{s}_d)^{\frac{1}{d}}}{\rho\Gamma(1+\frac{1}{d})}\left\langle \mathbf{P}_0^{\mathrm{flat}}\right\rangle = \frac{(\rho \mathbf{s}_d)^{\frac{1}{d}}}{\rho\Gamma(1+\frac{1}{d})}\left\langle \mathbf{F}_0^{\mathrm{flat}}\right\rangle. \tag{3.40}$$

The corresponding discrete, dimensionless spatial volumes $V_\pm$ are

$$V_+(\mathcal{A}_+) = \mathbf{s}_d{}^{\frac{1}{d}} \frac{1}{\Gamma(1+\frac{1}{d})} P_0, \quad V_-(\mathcal{A}_-) = \mathbf{s}_d{}^{\frac{1}{d}} \frac{1}{\Gamma(1+\frac{1}{d})} F_0. \tag{3.41}$$

Exercise 3.12 Show that the relationship Eq. (3.40) also holds for causal sets that faithfully embed into the flat spacetime $M \sim T^n \times I$ in the large ρ limit and with zero extrinsic curvature slices.

Next, we ask whether this formula generalises to all spacetimes regions $M \in \Sigma \times \mathbb{R}$ with Σ compact. As described above, we focus our attention on the GNN, $\mathcal{N}_\Sigma$ associated with Σ [118,119]. The flat spacetime result suggests that we again begin with the random variables $\mathbf{F}_0$ and $\mathbf{P}_0$ and first evaluate Eq. (3.38).

The integrals over $J^\pm(\Sigma)$ must be calculated inside $\mathcal{N}_\Sigma$, i.e., in $J^\pm(\Sigma) \cap \mathcal{N}_\Sigma$. Using the GNN coordinates of Eq. (3.36)

$$\int dV_p = \int d^{d-1}x \int d\tau \sqrt{h(\tau,\mathbf{x})}. \tag{3.42}$$

As an intermediate step in the evaluation we will use cut-offs $\pm\epsilon$ for the integrals over τ before taking the large ρ limit. Expanding $\sqrt{h(\tau,\mathbf{x})}$ in powers of τ

$$\sqrt{h(\tau,\mathbf{x})} = \sqrt{h(0,\mathbf{x})}\left(1 + \frac{1}{2}\frac{\dot{h}(0,\mathbf{x})}{h(0,\mathbf{x})}\tau + O(\tau^2)\right) = \sqrt{h(0,\mathbf{x})}\left(1 - K(0,\mathbf{x})\tau + O(\tau^2)\right), \tag{3.43}$$

where $K(0,\mathbf{x}) = -\frac{1}{2}\frac{\dot{h}(0,\mathbf{x})}{h(0,\mathbf{x})}$ is the trace of the extrinsic curvature at $(0,\mathbf{x})$.

In order to proceed further in evaluating Eq. (3.38) we also need to expand the volumes $V_\blacktriangle(\tau,\mathbf{x})$ and $V_\blacktriangledown(\tau,\mathbf{x})$ in terms of τ. While the small volume expansion of a causal diamond in an RNN [18,120,121] in terms of its proper time is well known, what is required here is not the volume of the diamond, but of the regions $\blacktriangle$ and $\blacktriangledown$ with the GNN coordinate τ as the expansion parameter. It is a somewhat lengthy but straightforward exercise to show that the RNN and GNN coordinates can be combined suitably around $(0,\mathbf{x})$ [118,119] to obtain a common expansion parameter which we will continue to call τ.

The corrections to the flat space volume of the causal diamond in the RNN are of order τ^2, and these come from the corrections to $\sqrt{-g(x)}$ as well as from a change in the shape of the lightcone boundaries of the diamond [121]. However, Eq. (3.43) already gives us a hint that in $\blacktriangle$ and $\blacktriangledown$, an order τ (and hence leading order) contribution comes from the extrinsic and intrinsic curvature of the portion of Σ that intersects the causal diamond. This leads to a considerable simplification for small enough τ,

$$V_{\blacktriangle}(\tau, \mathbf{x}) = \mathrm{s}_d \tau^d \left(1 + \frac{d}{2(d+1)} K(0, \mathbf{x}) \tau\right) + O(\tau^{d+2})$$
$$V_{\blacktriangledown}(-\tau, \mathbf{x}) = \mathrm{s}_d \tau^d \left(1 - \frac{d}{2(d+1)} K(0, \mathbf{x}) \tau\right) + O(\tau^{d+2}). \quad (3.44)$$

Expanding in powers of ρ Eq. (3.38) simplifies to

$$\langle \mathbf{P}_0 \rangle = \rho \Gamma\left(1 + \frac{1}{d}\right)\left(\rho \mathrm{s}_d\right)^{-\frac{1}{d}} \mathrm{vol}(\Sigma) - \frac{1}{2} d\, \rho \Gamma\left(2 + \frac{2}{d}\right)\left(\rho \mathrm{s}_d\right)^{-\frac{2}{d}} \int_\Sigma d^{d-1}x \sqrt{h} K + O\left(\rho^{1-\frac{3}{d}}\right),$$
$$\langle \mathbf{F}_0 \rangle = \rho \Gamma\left(1 + \frac{1}{d}\right)\left(\rho \mathrm{s}_d\right)^{-\frac{1}{d}} \mathrm{vol}(\Sigma) + \frac{1}{2} d\, \rho \Gamma\left(2 + \frac{2}{d}\right)\left(\rho \mathrm{s}_d\right)^{-\frac{2}{d}} \int_\Sigma d^{d-1}x \sqrt{h} K + O\left(\rho^{1-\frac{3}{d}}\right). \quad (3.45)$$

Thus,

$$\mathrm{vol}(\Sigma) = \lim_{\rho \to \infty} \frac{1}{2}\left(\rho \Gamma\left(1 + \frac{1}{d}\right)\right)^{-1}\left(\rho \mathrm{s}_d\right)^{\frac{1}{d}}\left(\langle \mathbf{F}_0 \rangle + \langle \mathbf{P}_0 \rangle\right). \quad (3.46)$$

As in the calculations of the number of chains C_k in an RNN in Sect. 3.3, we can extract more geometric information from the $\mathcal{A}_\pm$ than we might have initially guessed. In this case, we find that

$$\int_\Sigma d^{d-1}x \sqrt{h} K = \lim_{\rho \to \infty} 2\left(d\, \rho \Gamma\left(2 + \frac{2}{d}\right)\right)^{-1}\left(\rho \mathrm{s}_d\right)^{\frac{2}{d}}\left(\langle \mathbf{F}_0 \rangle - \langle \mathbf{P}_0 \rangle\right). \quad (3.47)$$

This is the well know Gibbons-Hawking-York (GHY) boundary term for the action! It is quite startling that the simplest of counting-type order invariants $|\mathcal{A}_\pm|$ contains the GHY boundary term for the Einstein-Hilbert action.

Geometrically, this term can be expressed as the normal derivative of the volume of Σ

$$S_{GHY}\left[\Sigma, M^\pm\right] = \mp \frac{1}{l_p^{d-2}} \int_\Sigma d^{d-1}x \sqrt{h}\, K = \mp \frac{1}{l_p^{d-2}} \frac{\partial}{\partial n} \int_\Sigma d^{d-1}x \sqrt{h} = \mp \frac{1}{l_p^{d-2}} \frac{\partial}{\partial n} \mathrm{vol}(\Sigma), \quad (3.48)$$

which, in the discrete setting, is given by the difference in the discrete volume as one crosses Σ.

In [118] a whole family of invariants P_k and F_k was defined. Just as P_0 counts the number of elements in C for which there are no other elements in $J^+(\Sigma) \cap I^-(\mathcal{A}_+)$, let P_k be the number of elements in C for which there are exactly k elements in $J^+(\Sigma) \cap I^-(\mathcal{A}_+)$. Similarly let F_k be the number of elements in C for which there are exactly k elements in $J^-(\Sigma) \cap I^+(\mathcal{A}_+)$. It can be shown that there are other combinations of the P_k and F_k that yield alternative expressions for vol(Σ) and $\int_\Sigma d^{d-1}x\,\sqrt{h}K$:

$$\mathrm{vol}(\Sigma) = \lim_{\rho\to\infty}\frac{1}{2}\left(\rho\,\Gamma\left(1+\frac{1}{d}\right)\right)^{-1}\left(\rho s_d\right)^{\frac{1}{d}}\left(\sum_k p_k\langle \mathbf{F}_k\rangle + \sum_k q_k\langle \mathbf{P}_k\rangle\right),$$

$$\int_\Sigma d^{d-1}x\,\sqrt{h}K = \lim_{\rho\to\infty} 2\left(d\,\rho\,\Gamma\left(2+\frac{2}{d}\right)\right)^{-1}\left(\rho s_d\right)^{\frac{2}{d}}\left(\sum_k p_k\langle \mathbf{F}_k\rangle + \sum_k q_k\langle \mathbf{P}_k\rangle\right), \tag{3.49}$$

where the strings $\{p_k\}$ and $\{q_k\}$ satisfy

$$\sum_k p_k\frac{1}{k!}\Gamma\left(\frac{1}{d}+k\right)+\sum_k q_k\frac{1}{k!}\Gamma\left(\frac{1}{d}+k\right)=0$$
$$\sum_k p_k\frac{1}{k!}\Gamma\left(\frac{2}{d}+k\right)-\sum_k q_k\frac{1}{k!}\Gamma\left(\frac{2}{d}+k\right)=1. \tag{3.50}$$

It is an open question, and worth investigating, whether including larger k terms in the expressions Eq. (3.49) leads to a better control of fluctuations.

Since an inextendible antichain $\mathcal{A} \subset C$ contains no relational information, it admits only the trivial, discrete topology. To recover non-trivial structure on $\mathcal{A}$ we have to "borrow" information from its collar neighbourhood and construct induced structures on $\mathcal{A}$.

For $e \in \mathrm{Fut}(\mathcal{A})$, the discrete proper time to $\mathcal{A}$ is the length of the longest chain $l(e, \mathcal{A})$ from e to $\mathcal{A}$. Additionally, we can define the discrete spacetime volume from e to $\mathcal{A}$ to be the cardinality $V(e, \mathcal{A}) = |\mathrm{IFut}(\mathcal{A}) \cap \mathrm{IPast}(e)|$ (see Eq. (2.16)) and the closed order interval $[e, e'] = \mathrm{IFut}(e) \cap \mathrm{IPast}(e')$.

Both these measures of distance from $\mathcal{A}$ can be used to define families of future collar neighbourhoods or "thickenings" of $\mathcal{A}$:

$$\mathcal{T}_l(\mathcal{A}) = \{e \in \mathrm{IFut}(\mathcal{A}) \mid l(\mathcal{A}, e) \le l\},$$
$$\mathcal{T}_K(\mathcal{A}) = \{e \in \mathrm{IFut}(\mathcal{A}) \mid |[\mathcal{A}, e]| \le K\}. \tag{3.51}$$

One can similarly define the past collar neighbourhoods of $\mathcal{A}$.

Exercise 3.13 Using a comparison between the different causal sets in Fig. 1.1, show that $\mathcal{T}_l(\mathcal{A})$ and $\mathcal{T}_K(\mathcal{A})$ are, in general, distinct subsets of C. Use the minimal or past-most antichain as $\mathcal{A}$.

In Sect. 3.3 we discussed the construction of a simplicial complex on $\mathcal{A}$ using the chain homology. By fixing l we therefore obtain an upper bound on the dimension of the simplices, since this bounds the length of the longest chain from $\mathcal{A}$. This is in stark contrast to the simplicial complexes obtained from triangulations of spacetime where the largest dimensions of the simplex is determined by the spacetime dimension.

In what follows we will focus our attention on $\mathcal{T}_K(\mathcal{A})$ since it provides a very useful induced structure on $\mathcal{A}$. The analogue of $\mathcal{T}_K(\mathcal{A})$ in the continuum is the future collar neighbourhood of a Cauchy hypersurface Σ

$$\mathcal{T}_V(\Sigma) \equiv \{p \in I^+(\Sigma) | \text{vol}(\Sigma, p) \leq V\}. \tag{3.52}$$

We choose V such that $\mathcal{T}_V(\Sigma)$ is itself contained in a GNN, $\mathcal{N}_\Sigma$ and will consider $\mathfrak{C}_\rho(\mathcal{T}_V(\Sigma))$, where $\mathcal{A}$ is the set of minimal elements of any $C \in \mathfrak{C}_\rho(\mathcal{T}_V(\Sigma))$. As before, for convenience of notation, we will drop the reference to the faithful embedding $\varphi : C \hookrightarrow_\rho \mathcal{T}_V(\Sigma))$ in what follows, i.e., C is taken to be the same as $\varphi(C)$. Let K denote the maximum discrete volume in C to $\mathcal{A}$, i.e., $K = sup_{e \in C} |[\mathcal{A}, e]|$. The index $k = \{0, \ldots, K\}$ will be used to denote the discrete thickening parameter, and $\mathfrak{v} \leq V$ the continuum thickening parameter.

Our first task is to borrow information from $\mathcal{T}_k(\mathcal{A})$ to obtain topological information about Σ, thus going beyond the trivial discrete topology of $\mathcal{A}$. For $e \in \mathcal{T}_k(\mathcal{A})$ define the **beam** $O(e) \equiv \text{IPast}(e) \cap \mathcal{A}$. The set of beams $\mathcal{O}_k(\mathcal{A}) = \{O_e | e \in \mathcal{T}_k(\mathcal{A})\}$ therefore provides a **discrete covering** of $\mathcal{A}$, i.e., $\mathcal{A} = \cup_{\mathcal{O}_k(\mathcal{A})} O$. The topology associated with this cover is however still discrete, but we can nevertheless use it to obtain "coarse" topological invariants.

Exercise 3.14 Show that the set of all $\mathcal{O}_k(\mathcal{A})$ with $k \geq 0$ generates the discrete (and hence trivial) topology on $\mathcal{A}$.

The discrete covering $\mathcal{O}_k(\mathcal{A})$ can be used to construct the **nerve simplicial complex** [122] of $\mathcal{A}$ as follows. Assign a vertex or 1-simplex to every beam $O \in \mathcal{O}_k(\mathcal{A})$, assign a 2-simplex, to every non-empty two-way intersection of beams $O_1 \cap O_2$, and in general, assign a j-simplex to every non-empty j-way intersection of beams (see Fig. 3.7.) This generates the nerve simplicial complex $\mathcal{N}(\mathcal{O}_k(\mathcal{A}))$ associated with $\mathcal{O}_k(\mathcal{A})$.

What information about Σ can we hope to glean from $\mathcal{N}(\mathcal{O}_k(\mathcal{A}))$? In the Riemannian space (Σ, h), the nerve simplicial complex associated with a locally finite open covering $\mathcal{O}(\Sigma)$ of *convex* sets was shown by de Rham and Weil to be homotopic to Σ [123,124]. An open set $O \subset \Sigma$ is said to be **convex** if every pair of points in the set can be joined by a unique geodesic lying wholly in that set, of arc-length less

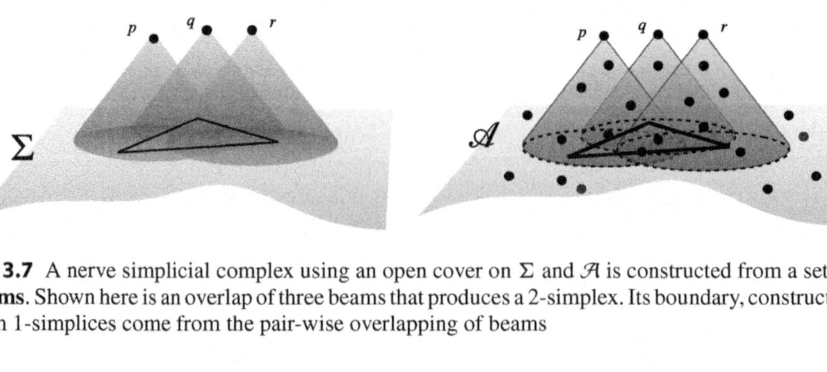

Fig. 3.7 A nerve simplicial complex using an open cover on Σ and $\mathcal{A}$ is constructed from a set of **beams**. Shown here is an overlap of three beams that produces a 2-simplex. Its boundary, constructed from 1-simplices come from the pair-wise overlapping of beams

than the "convexity radius" of the space. The **convexity radius** r_p of any $p \in \Sigma$ is given by the largest $r > 0$ such that the distance function $d(p, x)$ on Σ is a convex function in the open ball $B_r(p)$ and the convexity radius of Σ is the infimum over all r_p. The convexity condition ensures that the sets in the open cover are not too large and can therefore capture the local topological features of Σ. This need for convexity is most simply illustrated by the example of an open cover of $\mathbb{S}^1$ consisting of just two overlapping (and hence non-convex) sets. The associated nerve simplicial complex is thus just a 1-simplex, which is therefore not homotopic to $\mathbb{S}^1$. An obvious problem arises when extending this construction to Lorentzian spacetime, since there is no natural definition of a convexity radius that will "localise" the sets in a given Alexandrov interval covering, for example.

Let us continue further with the continuum before turning back to the discrete case. We can "foliate" $\mathcal{T}_V(\Sigma)$ (Eq. (3.52)) with spatial hypersurfaces $\Sigma_{\mathfrak{v}}$, $\mathfrak{v} < V$. The collection of beams $O_{\mathfrak{v}}(\Sigma) \equiv \{I^-(p) \cap \Sigma | p \in \Sigma_{\mathfrak{v}}\}$ from $\Sigma_{\mathfrak{v}}$ to Σ thus forms an open covering of Σ. For every Σ there is a **convexity volume** $\mathfrak{v}_\Sigma$ such that for all $\mathfrak{v} < \mathfrak{v}_\Sigma$ the beams in the open cover $O_{\mathfrak{v}}(\Sigma)$ are convex – such a non-trivial infimum is guaranteed for a compact Σ. A locally finite subcover can then be extracted from $O_{\mathfrak{v}}$ to construct a nerve simplicial complex $\mathcal{N}(O_{\mathfrak{v}}(\Sigma))$, which the de Rham-Weil theorems guarantee is homotopic to Σ.

It is evident that $O_k(\mathcal{A})$ is the causal set analog of $O_{\mathfrak{v}}(\Sigma)$, which is naturally locally finite. The idea is therefore to see if $\mathcal{N}(O_k(\mathcal{A}))$ is homological to $\mathcal{N}(O_{\mathfrak{v}}(\Sigma))$. An important feature of $\mathcal{N}(O_k(\mathcal{A}))$ is that there is an "evolution" in the homology groups as the discreteness volume k increases. To begin with, $O_{k=0}$ consists of just the elements of $\mathcal{A}$. Thus $\mathcal{N}(O_{k=0})$ is a set of $|\mathcal{A}|$ unconnected vertices with $\mathbb{H}_0(k=0) = \bigoplus_{|\mathcal{A}|} \mathbb{Z}$, $\mathbb{H}_i = 0, i \neq 0$.

Next, consider $\mathcal{T}_{k=1}$. Every $e \in \mathcal{T}_{k=1} \backslash \mathcal{A}$ is such that $|\mathrm{Past}(e)| = 1$, i.e., it is at most related to a single element in $\mathcal{A}$. In a sprinkling, for $\mathfrak{v}$ close to the discreteness scale, there are large fluctuations and hence $\mathcal{T}_{k=1}$ will typically only include a few more points to the future of $\mathcal{A}$ without vastly improving connectivity. Thus, one still has a nerve simplicial complex that is very disconnected corresponding to a large $\mathbb{H}_0(k=1)$, but which is typically of smaller rank than that at $k = 0$. In order to achieve sufficient connectivity, we need $k \gg 1$, i.e., $\mathfrak{v} \gg \rho^{-1}$. However, we also require that $\mathfrak{v} < \mathfrak{v}_\Sigma$, the convexity volume of Σ which sets a "meso-scale", since otherwise the de Rham-Weil results are no longer valid. In this range of k we should therefore see a transition from the highly disconnected nerve simplex to one that is homological to Σ. Indeed, one cannot expect more than homological equivalence since homotopic equivalence is not applicable in such a discrete setting.

It was shown in [34] that for any globally hyperbolic spacetime (M, g) with compact Cauchy hypersurface, the $\mathcal{N}(O_k(\mathcal{A}))$ associated with an inextendible antichain $\mathcal{A}$ is homological to Σ when there is a clear "separation of scales" $\mathfrak{v}_\Sigma > \mathfrak{v} >> \rho^{-1}$. The argument uses compactness to show that the possibility of "voids" in the coverings is highly suppressed and hence the resulting complex is faithful to the continuum.

Exercise 3.15 Calculate the homology groups for the nerve simplicial complex for different k for (a) the crown poset $\mathbf{cr}_{2n}$ with $n = 4$ and (b) the tower with two $n = 4$

crown posets stacked one on top of the other. Use the minimal (pastmost) antichain as $\mathcal{A}$ as well as a non-minimal one of cardinality less than 4.

Numerical simulations in $d = 2, 3$ verify these arguments [34, 125]. The evolution of the homology groups with thickening parameter k moreover shows that the rank of $\mathbb{H}_0$ initially decreases rapidly and stabilises to one (for a connected Σ) while the rank of the relevant $\mathbb{H}_i$ increases until it also stabilises to the continuum values [125]. This "persistent" or "stable" homology is useful to pick up continuumlike properties in a causal set.

The construction above, and the deviations we have seen from flatness in the calculations of F_0 and P_0, suggest that further geometric information can be ferreted out of $\mathcal{T}_k(\mathcal{A})$.

Again, we turn to the continuum and construct a (piece-wise linear) estimate of the induced distance function on Σ using $\mathcal{T}_V(\Sigma)$. Before we do so, it is useful to consider an example on the $d = 2$ Euclidean plane $\mathbb{R}^2$. Let $\gamma : [0, 1] \to \mathbb{R}^2$ be a smooth curve, with $\gamma(0) = p$, $\gamma(1) = q$. Imagine that we only have access to the flat geometry of $\mathbb{R}^2$, and use only this to construct an approximate pre-distance function $d(p, q)$ on γ. We begin by chopping γ into a finite set of segments $\{\gamma_i\}, i = 1, \ldots m$ with end points w_i, where $w_0 = p$, $w_{m+1} = q$ and approximating the length of each γ_i by the linear distance between w_i and w_{i+1} in $\mathbb{R}^2$. This in turn can be related to the suspended area $A_{i,i+1}$ between the linear distance and curve as shown in Fig. 3.8, where $d(w_i, w_{i+1}) \approx \sqrt{\frac{8A_{i,i+1}}{\pi}}$ for a small arc. This gives a piecewise linearisation of γ which improves as one increases the number of segments m systematically. The length of γ can then be found by taking the shortest distance over all such pre-distances. However, if the end points are too close to each other in $\mathbb{R}^2$ compared to the curvature of γ, this minimisation procedure will give a very poor estimate, unless we impose the further constraint that the segments should be short compared to the curvature scale of γ. This simple example is useful in navigating the Lorentzian case below.

Let us begin with the simplest Lorentzian geometry, $\mathbb{M}^d$, with a flat spatial hypersurface Σ at coordinate time $t = 0$. For any $r \in I^+(\Sigma)$, the distance $\mathbf{d}$

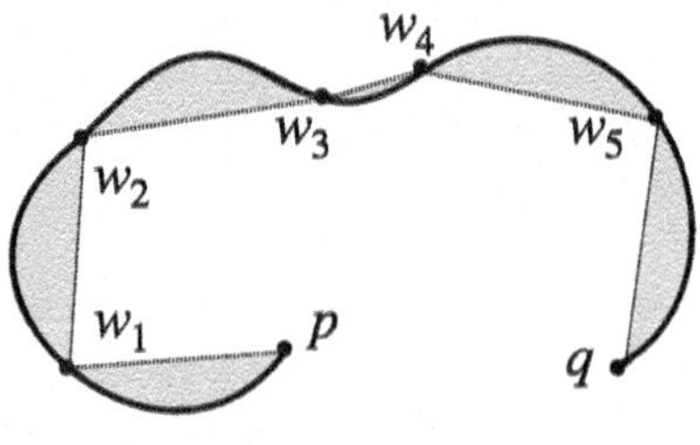

Fig. 3.8 An example of a piece-wise linearisation of a curve γ from p to q in $\mathbb{R}^2$. The approximate distance along γ can be calculated using the square root of the suspended area – the smaller the area, the better the approximation

between diametrically opposite points in its beam $B(r)_\Sigma \equiv I^-(r) \cap \Sigma$ is related to $V(r) \equiv \text{vol}(\Sigma, r)$ by

$$\text{vol}(r) = \zeta_d \left(\frac{\mathbf{d}}{2}\right)^d, \tag{3.53}$$

where ζ_d is defined as before. Conversely, if we start with a pair of points $p, q \in \Sigma$, there is a unique $\mathbf{r}_m$ in their common causal future $J^+(p) \cap J^+(q)$ which minimises the volume $\text{vol}(\mathbf{r}_m)$. Hence the distance in Σ is given by

$$\mathbf{d}(p,q) = 2\left(\frac{\text{vol}(\mathbf{r}_m)}{\zeta_d}\right)^{\frac{1}{d}}. \tag{3.54}$$

It is important to emphasise that this construction uses only the causal structure and the spacetime volume to obtain the exact induced spatial distance on Σ. We have used the simplicity of the flat geometries involved – both spacetime and spatial – to get an exact result. We now generalise the construction to a GNN $\mathcal{N}_\Sigma$, where $p, q \in \Sigma$ lie in an RNN $\mathcal{N} \subset \mathcal{N}_\Sigma$. We will also require that $p, q \in \mathcal{O}$, a RNN within Σ, such that $D(\mathcal{O}) \subset \mathcal{N}$, where $D(X)$ denotes the domain of dependence of X (see [82, 126].)

In such a neighbourhood, and for any $r \in I^+(\Sigma) \cap \mathcal{N}$, $\text{vol}(r)$ is of the form $V_\blacktriangle$ given in Eq. (3.44). Since $\mathcal{O}$ itself is an RNN in Σ and $D(\mathcal{O})$ lies in an RNN of the spacetime, which is diffeomorphic to an open ball in $\mathbb{R}^d$, there exists an $\mathbf{r}_m \in J^+(p) \cap J^+(q)$ for which $\text{vol}(\mathbf{r}_m)$ is minimised. In analogy with the flat spacetime case, we define the **pre-distance function**

$$\widetilde{d}(p,q) \equiv 2\left(\frac{\text{vol}(\mathbf{r}_m)}{\zeta_d}\right)^{\frac{1}{d}}. \tag{3.55}$$

Using Eq. (3.44) this reduces to

$$\widetilde{d}(p,q) = 2\tau\left(1 + \frac{1}{2(d+1)}K\tau\right) + O(\tau^2), \tag{3.56}$$

where τ is the proper distance from $\mathbf{r}_m$ to Σ. As p and q get closer to each other, τ becomes smaller, and the approximation to the flat spacetime distance improves. Just as in the calculation of the homology of Σ, the combined GNN and the RNN also sets a maximum scale or "meso-scale" (similar to the convexity radius) which we call δ and we will require that $0 < \text{vol}(\mathbf{r}_m) \le \delta$, hold for Eq. (3.56).

In order to define $\widetilde{d}$ over all of Σ, one must then construct curves from these δ-patches. For every p, q for which $\text{vol}(\mathbf{r}_m) > \delta$, construct a δ-piece-wise linear curve $\gamma(p,q)$, between $(s_0 = p, s_1, s_2, \ldots, s_k, s_{k+1} = q)$ where $\widetilde{d}(s_i, s_{i+1}) \le \delta$. Such a curve is of course not unique. But we can use it to define

$$\widetilde{d}_\gamma(p,q) \equiv \sum_i \widetilde{d}(s_i, s_{i+1}). \tag{3.57}$$

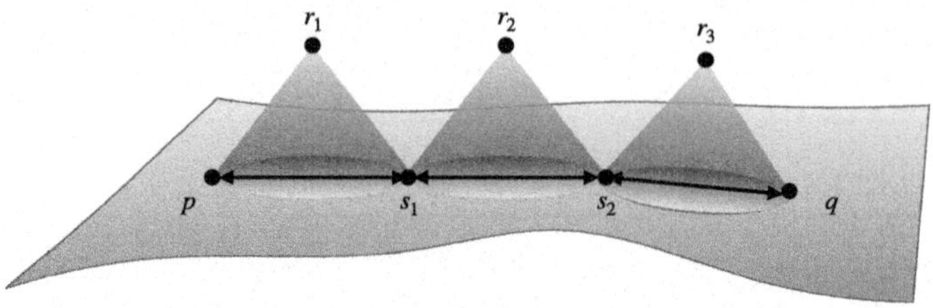

Fig. 3.9 An example of a piece-wise linear curve γ from p to q, maximally constrained by the size of the beam

Figure 3.9 shows an example of this construction. Varying over all curves, as for the curve in $\mathbb{R}^2$, gives us a distance function which is itself an estimate of the induced distance function.

A simple geometry experiment (real or imagined) that illustrates this idea quite nicely is the following. For the experimentally inclined, take a set of paper cones of the same angle (the kind peanuts are wrapped in by street vendors in India), but of varying heights. We can place the base of a cone on a curved surface Σ (a pillow will do) to approximate its intrinsic distance. For any pair p, q on the surface choose a cone for which p and q are at a diametrically oppose ends of the base. The height (or equivalently the cube root of the volume of the cone) serves as a proxy for the induced pre-distance between p and q. The curvature scale of Σ (both intrinsic and extrinsic) limits the accuracy of the induced pre-distance: when the height is small with respect to the curvature scale, the estimate is fairly good, but as it becomes larger (i.e., with points that are further apart) the induced distance is poorly estimated. Imposing a cut-off on the height, determined by the smallest curvature scale on Σ, improves the induced pre-distance estimate. The estimated induced distance, obtained by minimising over the pre-distance of piece-wise linear curves therefore depends on this cut-off.

The construction carries over in a straightforward manner to the causal set case. Let $C \sim_\rho (M, g)$, where (M, g) is a globally hyperbolic spacetime, and let $\mathcal{A}$ denote an inextendible antichain in C. Define $n(e) \equiv |\mathrm{IFut}(\mathcal{A}) \cap \mathrm{IPast}(e)|$ for $e \in \mathrm{Fut}(\mathcal{A})$. For any pair $e_1, e_2 \in \mathcal{A}$ with a non-empty common future, i.e., $\mathrm{Fut}(e_1) \cap \mathrm{Fut}(e_2) \neq \emptyset$, there is an $e_{12} \in \mathrm{Fut}(e_1) \cap \mathrm{Fut}(e_2)$ which minimises $n(e)$. Using the analogy with the continuum, define a pre-distance function on $\mathcal{A}$ as

$$\widetilde{\mathbf{d}}(e_1, e_2) \equiv 2\left(\frac{n(e_{12})}{\zeta_d}\right)^{\frac{1}{d}}. \tag{3.58}$$

When $n(e_{12}) \sim 1$ or equivalently $\mathrm{vol}(e_{12}) \equiv \mathrm{vol}(\mathcal{A}, e_{12}) \simeq \rho^{-1}$, the fluctuations are large and this increases the error in $\widetilde{\mathbf{d}}$ considerably. This results in a much larger distance function on *small* scales than expected in the continuum. This feature suggests there is a **discrete asymptotic silence** [127, 128]: if one were to take spatial constructions seriously, then it would appear that light cones are effectively narrower near the discreteness scale.

Let $C \sim_\rho (\mathcal{N}_\Sigma, g)$ and consider its restriction to the region $\mathcal{T}_V(\Sigma) \subset \mathcal{N}_\Sigma$ assuming that $\rho V >> 1$. Let $\mathcal{A}$ denote the minimal (past-most) antichain in C. Thus, for every pair $e_1, e_2 \in C$, if there is an element e_{12} in their common future in C, then $\mathrm{vol}(e_{12}) < V$. Thus, as in the continuum, the "meso-scale" V sets an upper bound on the pre-distance function. We define the k-element ordered spatial path set $\chi_k(p, q) \equiv (\mathbf{w}_0, \mathbf{w}_1, \ldots, \mathbf{w}_k)$ with distinct $\mathbf{w}_i \in \mathcal{A}$ and $\mathbf{w}_0 = p$, $\mathbf{w}_k = q$ and $\widetilde{\mathbf{d}}(\mathbf{w}_i, \mathbf{w}_{i+1}) \leq \ell$. Let $\mathcal{W}(p, q)$ denote the set of all such spatial paths from p to q in $\mathcal{A}$. Just as in the continuum, we define the spatial path distance along χ_k to be

$$\widetilde{\mathbf{d}}_{\chi_k}(p, q) \equiv \sum_{i=0}^{k-1} \widetilde{\mathbf{d}}(\mathbf{w}_i, \mathbf{w}_{i+1}). \tag{3.59}$$

Minimising over all the spatial paths gives us the distance function associated with the mesoscale ℓ

$$\mathbf{d}_\ell(a, b) \equiv \inf_{\chi \in \mathcal{W}} \widetilde{\mathbf{d}}_\chi(a, b). \tag{3.60}$$

As long as $\rho V >> 1$, $\mathbf{d}_\ell(p, q)$ is a good measure of the distance function on Σ. This is confirmed by numerical simulations [128], an example of which is shown in Fig. 3.10. While there is a strong convergence to the continuum at large distances, there is a stark deviation from the continuum at small distances. The latter is a sign of **asymptotic silence** at scales close to the discreteness scale. It roughly means that a spatial reconstruction of the continuum at scales of the order of the discreteness scale corresponds to a continuum approximation in which light cones narrow down.

The distance function can then be used to obtain a spectral dimension on $\mathcal{A}$. Spectral dimensions are useful in teasing out geometric information in various approaches

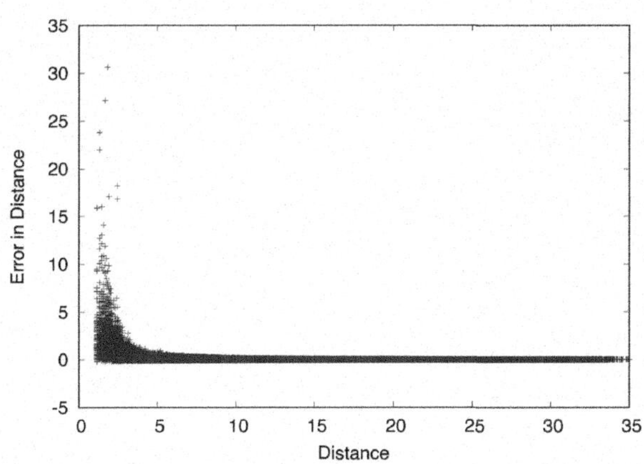

Fig. 3.10 The predistance function on a constant time hypersurface can be calculated for a sprinkling into a region of $\mathbb{M}^3$. The graph shows the error in the predistance function with respect to the continuum distance as function of the predistance function. At small distances, the error is very large and is an indication of "asymptotic silence", while at larger distances, the error becomes negligible

to quantum gravity, for example in $4d$ CDTs, where a short range dimension of 2 was discovered [3] which therefore differs from the Hausdorff dimension. In causal sets there have been attempts to extract such a spectral dimension [129], but the nonlocality makes this difficult. Using the distance function on $\mathcal{A}$ one can extract a spatial spectral dimension from $\mathcal{A}$ which agrees with the Hausdorff dimension at intermediate scales but leads to a dimensional reduction at smaller scales [130].

There is more spatial information we can obtain from $\mathcal{A}$. The distance function gives rise to a complete weighted graph $\mathcal{G}_\ell(\mathcal{A})$ where each vertex is a point of $\mathcal{A}$ and each edge is weighted by the distance function d_ℓ. This can be used to extract more geometric information about Σ, for example, spatial curvature estimates [131].

3.5 Order Intervals

In a spacetime that is strongly causal, the topology generated by the Alexandrov intervals $I(p, q)$ is the manifold topology [83]. Since each $I(p, q)$ is defined using the causal structure, and has an associated spacetime volume, it is a natural to use the discrete version, namely, the order interval (e, e') to construct order invariants.

The intersection of two Alexandrov intervals in $\mathbb{M}^d$ can be of arbitrarily small volume (for example, very close to the discreteness scale) without making the Alexandrov topology trivial. As we pointed out earlier, it is the same as the manifold topology. Can order intervals for $C \in \mathfrak{C}_\rho(\mathbb{M}^d)$ be used similarly? Because of non-local discreteness, almost surely every element in C lies in an order interval (e, e') in which it is the sole element, which means that the order intervals generate the discrete (trivial) topology. Instead, it is more useful to obtain the coarse topology of the spacetime in the form of topological invariants as we have done in Sect. 3.4. There, the homology of a spatial hypersurface was obtained from the underlying causal set using an induced covering of the corresponding inextendible antichain. A natural extension of this construction is to use order intervals as a cover of C and construct its nerve simplicial complex.

In Fig. 3.11 we show such an example of a partial covering for a $d = 2$ cylinder spacetime $ds^2 = -dt^2 + d\theta^2$. Again, as in the construction in Sect. 3.4, a constraint on the size of the intervals is necessary to exclude excessively "large" open sets. In the Riemannian case the de Rham-Weil construction requires that the open sets in the cover are convex, which needs the sets to be "small enough" with respect to the curvature scale. However, because of the Lorentzian character of the spacetime volume, even small Alexandrov intervals can span a "long" region of a spacetime. Figure 3.11 shows what can go wrong with the nerve simplicial complex construction in this case. The figure on the left shows an example of a covering by "fat" Alexandrov intervals of sufficiently small volume V. A covering by such sets will give the correct homology group for the cylinder, i.e., $\mathbb{H}_0 = \mathbb{Z}$, $\mathbb{H}_1 = \mathbb{Z}$ and $\mathbb{H}_i = 0, \forall i > 1$. However, fixing the volume V is not enough to ensure this. The figure on the right shows what can happen when the volume V intervals are not constrained to be "fat" with respect to the frame of the cylinder (remember that Lorentz invariance is broken in this geometry) and therefore need not be small with respect to the curvature or

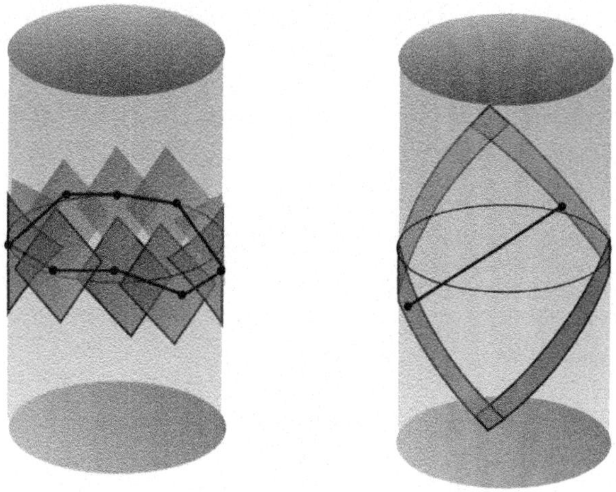

Fig. 3.11 A covering of a region in a $d = 2$ cylinder spacetime using Alexandrov intervals. In the figure on the left, we have chosen Alexandrov intervals which have volume V but are also "fat" in the frame shown. On the right we show a pair of intersecting "long and skinny" Alexandrov intervals which wrap around the cylinder but are of the same volume V

topological scale. In this case, there can be two intervals that wrap around the cylinder and hence intersect at two disjoint regions. This leads to the trivial homology $\mathbb{H}_0 = \mathbb{Z}, \mathbb{H}_i = 0, \forall i \geq 1$. Thus, limiting to small V is not sufficient to constrain the nature of the cover.

An additional restriction to a "small enough" slab of the spacetime is therefore required, similar to that used for the spatial homology. Such a spacetime covering has the advantage of being able to capture the homology of a topology changing region of a spacetime. It is an open question how well this discrete spacetime homology compares with the continuum.

Defining the size of an interval (e_i, e_j) to be its cardinality, we can construct a new family of order invariants in C by counting the number of k-element order intervals N_k in C. We will see how these simple order invariants can be used to define a discrete D'Alembertian operator, which in turn gives us an expression for the discrete Einstein-Hilbert or Benincasa-Dowker-Glaser action!

The simplest of these order intervals is already familiar to us – it is the number N_0 of zero size order intervals which is also the number of pairs of "nearest neighbours" or links $e \prec\!\!* \, e'$ in C.

We begin with an Alexandrov interval in $\mathbb{M}^d$, $I(p, q) = \mathbb{D}^d \subset \mathbb{M}^d$ with $\mathrm{vol}(p, q) = V$ and calculate $\langle \mathbf{N}_k \rangle$ for $\mathfrak{C}_\rho(\mathbb{D}^d)$ [132,133]. Let $\varphi : C \hookrightarrow_\rho \mathbb{D}^d$. Given $x \in \varphi(C)$, the probability for there to be a $y \in \varphi(C)$ such that $x \prec\!\!* \, y$, is given by the joint probability

$$P_{x \prec\!* y} = \mathrm{d}V_y \; e^{-\rho \mathrm{vol}(x,y)}. \tag{3.61}$$

Integrating over $x \in \mathbb{D}^d$ and $y \in J^+(x) \cap \mathbb{D}^d$ gives us the expectation value of the number of links in $\mathfrak{C}_\rho(\mathbb{D}^d)$

$$\langle \mathbf{N}_0 \rangle = \rho^2 \int_{\mathbb{D}^d} \mathrm{d}V_x \int_{J^+(x) \cap \mathbb{D}^d} \mathrm{d}V_y e^{-\rho \mathrm{vol}(x,y)}. \tag{3.62}$$

As in the calculation of the C_k in Sect. 3.3, we can use the Lorentz invariance of the intervals to choose convenient coordinates for the calculation. We begin with the integral over $I(x,q) = I^+(x) \cap \mathbb{D}^d$ and take x to be at the origin, i.e., $x = (0, \ldots, 0)$, with $q = (\tau, 0, \ldots, 0)$. Lorentz covariance implies that this integration depends only the proper time $\tau(x,q)$ and $\mathrm{vol}(x,q)$ of $I(x,q)$. Using lightcone coordinates

$$v = \frac{1}{\sqrt{2}}(t+r), \qquad u = \frac{1}{\sqrt{2}}(t-r), \tag{3.63}$$

we evaluate the first integral by using the binomial and power series forms [132]

$$\begin{aligned}
\int_{I(x,q)} \mathrm{d}V_y e^{-\rho \mathrm{vol}(x,y)} &= d(d-1)\zeta_d \int_0^{\tau/\sqrt{2}} \mathrm{d}v_y \int_0^{v} \mathrm{d}u_y (v_y - u_y)^{d-2} e^{-\rho\zeta_d (u_y v_y)^{d/2}} \\
&= d(d-1)\zeta_d \sum_{n=0}^{\infty} \frac{(-\rho\zeta_d)^n \, \tau^{d(n+1)}}{\sqrt{2}^{d(n+1)} d(n+1)n!} \sum_{k=0}^{d-2} \binom{d-2}{k} \frac{(-1)^k}{\frac{dn}{2}+k+1} \\
&= \zeta_d \Gamma(d+1) \sum_{n=0}^{\infty} \frac{(-\rho\zeta_d)^n \, \tau^{d(n+1)}}{\sqrt{2}^{d(n+1)} d(n+1)n!} \frac{\Gamma\left(\frac{dn}{2}+1\right)}{\Gamma\left(\frac{d}{2}(n+2)\right)},
\end{aligned} \tag{3.64}$$

where we have used the identity

$$\sum_{k=0}^{d-2} \binom{d-2}{k} \frac{(-1)^k}{\frac{dn}{2}+k+1} = \frac{\Gamma(d-1)\,\Gamma\left(\frac{dn}{2}+1\right)}{\Gamma\left(\frac{d}{2}(n+2)\right)}. \tag{3.65}$$

This gives us the analytic expression

$$\langle \mathbf{N}_0(\rho, V)\rangle = \Gamma(d)^2 \sum_{n=0}^{\infty} \frac{(-\rho V)^{n+2}}{(n+2)!} \frac{\Gamma\left(\frac{dn}{2}+1\right)}{\Gamma\left(\frac{d}{2}(n+2)\right)} \frac{\Gamma\left(\frac{d}{2}(n+1)+1\right)}{\Gamma\left(\frac{d}{2}(n+3)\right)}. \tag{3.66}$$

The remaining $\langle \mathbf{N}_k(\rho, V)\rangle$ can be obtained using

$$\begin{aligned}
\langle \mathbf{N}_k(\rho, V)\rangle &= \rho^2 \int_{I(p,q)} \mathrm{d}V_x \int_{I(x,q)} \mathrm{d}V_y \frac{(\rho V)^k}{k!} e^{-\rho V} \\
&= \frac{(-\rho)^{k+2}}{k!} \frac{\partial^k}{\partial \rho^k} \int_{I(p,q)} \mathrm{d}V_x \int_{I(x,q)} \mathrm{d}v_y \, e^{-\rho V} \\
&= \frac{(-\rho)^{k+2}}{k!} \frac{\partial^k}{\partial \rho^k} \rho^{-2} \langle \mathbf{N}_0(\rho, V)\rangle.
\end{aligned} \tag{3.67}$$

This can be rewritten in terms of generalised hypergeometric functions as

$$\langle \mathbf{N}_k(\rho, V)\rangle = \frac{(\rho V)^{k+2}}{(k+2)!} \frac{\Gamma(d)^2}{\left(\frac{d}{2}(k+1)+1\right)_{d-1}} \frac{1}{\left(\frac{d}{2}k+1\right)_{d-1}}$$
$$ {}_dF_d\left(\begin{array}{c} 1+k, \frac{2}{d}+k, \frac{4}{d}+k, \ldots, \frac{2(d-1)}{d}+k) \\ 3+k, \frac{2}{d}+k+2, \frac{4}{d}+k+2, \ldots, \frac{2(d-1)}{d}+k+2 \end{array} \middle| -\rho V\right), \quad (3.68)$$

where $(a)_n$ is the Pochhammer symbol [132,133].

For large $n = \rho V$ it can be shown that the ratio

$$\frac{\langle \mathbf{N}_k(\rho, V)\rangle}{\langle \mathbf{N}_0\ (\rho, V)\rangle} = \frac{\Gamma(\frac{2}{d}+k)}{\Gamma(\frac{2}{d})\Gamma(k+1)} + o(\rho V) \quad (3.69)$$

where the order of the sub-leading terms depends on the spacetime dimension. Inverting this gives us a new flat spacetime **dimension estimator** [132, 133].

Simulations with a single causal set (with $n \geq 50$) shows a considerable agreement with the ensemble average and hence with the continuum calculation. This suggests that counting order intervals of a given size N_k over a large enough finite causal set leads to a "self-averaging", which suppresses fluctuations. In other words, counting N_k in a *single* finite causal set of large enough n is sufficient to determine whether it is approximated by Minkowski spacetime of a given dimension or not.

What we will now see is that the N_k contain *far* more geometric information than just the flat spacetime dimension. As shown in [36,37,134], they are the main ingredients for constructing the discrete flat spacetime D'Alembertian, and whose curved spacetime avatar gives the discrete Einstein-Hilbert action or the Benincasa-Dowker-Glaser action. We review in some detail the various ingredients that go into this very significant development.

For a causal set that is approximated by $\mathbb{M}^d$, every element has an infinite number of nearest neighbours which are linked to it, either to its past or to its future. As we have mentioned earlier, unlike the various regular latticisations of spacetimes, the valency of the causal set graph is infinite. Importantly, this means that there are no intrinsic directions that can be assigned to an element, since it lacks a meaningfully defined tangent space. At first sight this is troubling, since all continuum definitions of geometry are intimately linked to tensorial structures. However, if we remember that only scalars are truly covariant, we can start reexamining Lorentzian geometry from this very different perspective.

Equation (3.33) is an expression for the retarded or advanced Green function $K^{(2)}(e, e')$ of a scalar field on $C \in \mathfrak{C}_\rho(\mathbb{M}^2)$ using the causal matrix M_C. The continuum identity $\Box G(x, x') = \delta(x, x')$, suggests a candidate discrete scalar field D'Alembertian operator $B(e, e')$, as the "inverse" of the symmetrised $K^{(2)}(e, e')$, if it exists. As defined, the causal matrix M_C is singular, but can be symmetrised, along with the the **reflexive condition** $x \prec x$, which replaces the 0's on the diagonal with 1's. However, as shown in [135, 136], this is not always easy to implement since often there are "nearly singular" cases.

In [134] Sorkin instead constructed a scalar field D'Alembertian from the set of nearest neighbours of a given element in a causal set, in analogy with the standard construction on a lattice. Let $\phi : C \to \mathbb{R}$ be a real scalar field on a causal set $C \in \mathfrak{C}_\rho(\mathbb{M}^2)$. The causal past or future of any $e \in C$ can be divided into the set of nearest neighbours, the next nearest neighbours, and so on. Denote the set of nearest neighbours to the past by $L_0^-(e) \equiv \{e' \in C | e' \prec\!\!* \, e\}$ and those to the future by $L_0^+(e) \equiv \{e' \in C | e \prec\!\!* \, e'\}$. The size of the interval between e and any element in this set is therefore 0. The set of next to nearest neighbours are those whose interval with e is of cardinality 1 and splits into the sets $L_1^-(e) = \{e' \in C || I(e', e)| = 1\}$, $L_1^+(e) = \{e' \in C || I(e, e')| = 1\}$. Generalising, we define the set of kth nearest neighbours to e as $L_k^-(e) = \{e' \in C || I(e', e)| = k\}$, $L_k^+(e) = \{e' \in C || I(e, e')| = k\}$. This "layers" the past and future of e, into disjoint sets $L_k^\pm(e)$ for $k \geq 0$, as shown in Fig. 3.12. For $\mathfrak{C}_\rho(\mathbb{M}^d)$ in particular, the layers are roughly separated by the spacings between invariant hyperbolae at multiples of the discreteness scale.

Let us first revisit the question of how nonlocality prevents us from giving a reasonable definition of the first derivative of the ϕ for a causal set $C \in \mathfrak{C}_\rho(\mathbb{M}^2)$. For any $e' \in L_0^-(e)$, for example, we can define a candidate derivative in the $e' - e$ "direction" as $D_{e'e}\phi(e) \equiv \rho^{\frac{1}{d}}(\phi(e') - \phi(e))$, since they are roughly a distance $\rho^{-\frac{1}{d}}$ apart. For this to be well defined, we would require that as ρ increases, the difference $(\phi(e') - \phi(e))$ decreases and $D_{e'e}\phi(e)$ stabilises to a fixed value. This is however true *only* if $\phi(e)$ is approximately constant throughout C and not otherwise, for the following simple reason. Consider a frame in which $\phi(x)$ is a slowly varying function with respect to the discreteness scale. By this we mean that there exists a $V \gg \rho^{-1}$ such that, for every $\epsilon > 0$ and $x \in \mathbb{M}^d$, there exists a neighbourhood $O_x \in x$ with $\mathrm{vol}(O_x) > V$ such that $|\phi(x') - \phi(x)| < \epsilon$ for all $x' \in O_x$. Since $e' \prec\!\!* \, e$ can lie anywhere in the past, roughly in a region a proper time $\rho^{-\frac{1}{d}}$ away from e, it will likely lie outside the slowly varying region around e and hence $|\phi(e) - \phi(e')|$ can be very large even though e and e' are nearest neighbours. This is because the field configuration need not itself be Lorentz invariant. Increasing the sprinkling density ρ does not help resolve this fundamental issue.

On the other hand, almost miraculously, these "far away" neighbours do not contribute to the D'Alembertian as shown by Sorkin [134] for $C \sim_\rho \mathbb{M}^2$. On the lightcone lattice in $\mathbb{M}^2$ the D'Alembertian can be defined by taking the diamond plaquette to the immediate past of e, with vertices $\{e, e^u, e^v, e^{uv}\}$. A natural candidate for a discrete D'Alembertian is

$$-B\phi(e) = \phi(e) + \phi(e^{uv}) - \phi(e^u) - \phi(e^v). \tag{3.70}$$

In a causal set, this "irreducible" causal diamond can be realised in multiple ways, for example, with $e^u, e^v \in L_0^-(e)$, and $e^{uv} \in L_2^-(e)$. In the frame in which the field is slowly varying, if e^{uv} lies "far down" the hyperboloid and hence outside the slowly varying region, then simultaneously e^u is brought into the slowly varying region O_e and similarly, e^v becomes close to e^{uv}. Since these come with opposite signs, this

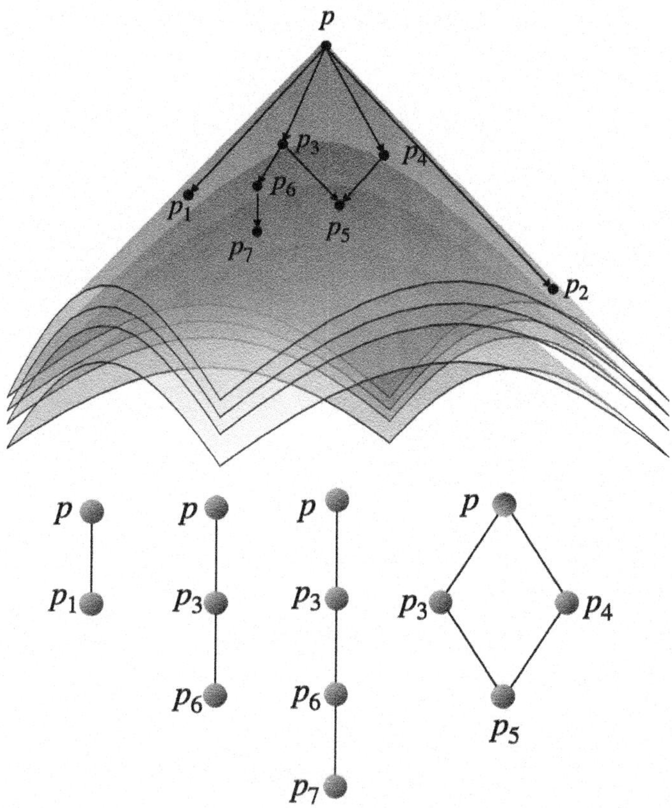

Fig. 3.12 **a** The layered past of an element $p \in C$ for $C \sim_\rho \mathbb{M}^3$ where each non-intersecting layer is $L_k^-(p)$. Here we show the layers $L_0^-(p)$, $L_1^-(p)$ and $L_2^-(p)$, which roughly form a hyperbola stretching all the way down the lightcone, with the elements $p_1, p_2, p_3, p_4 \in L_0^-(p)$, $p_6 \in L_1^-(p)$ and $p_5, p_7 \in L_2^-(p)$. **b** Examples of sets that contribute to N_0, N_1 and N_2. For N_2 we see the two possible types of intervals, which come from the number of possible 2-element causal sets. These are represented in **a** by the intervals $[p_5, p]$ and $[p_7, p]$

means that $-B\phi(e) \approx 0$ for these narrow causal diamonds. The only contribution is then from an effectively local causal diamond.

The discussion above is motivated by the diamond lattice, for which there is no layering of the past set. One therefore has to try something different for the causal set D'Alembertian, by playing around with the different layers to see which minimal combination of layers gives the D'Alembertian in the continuum limit. The following ansatz was proposed by Sorkin for the causal set D'Alembertian in $\mathbb{M}^2$:

$$\hat{B}^{(2)}\phi(e_i) \equiv B_{ij}^{(2)}\phi_j = 4\rho\left(-\frac{1}{2}\phi_i + \sum_{e_j \in L_0^-(e_i)} \phi_j - 2\sum_{e_j \in L_1^-(e_i)} \phi_j + \sum_{e_j \in L_2^-(e_i)} \phi_j\right), \tag{3.71}$$

where we've used the short-hand $\phi_i \equiv \phi(e_i)$. $B^{(2)}_{ij}$ is therefore a matrix given by

$$\frac{1}{4\rho}B^{(2)}_{ij} \equiv \begin{cases} -\frac{1}{2} & \text{for } i = j \\ 1, -2, 1 & \text{for } e_j \in L_0^-(e_i), L_1^-(e_i), L_2^-(e_i), \text{ respectively} \\ 0 & \text{otherwise} \end{cases} \tag{3.72}$$

We now follow the exposition of [137] to show that the expectation value of the associated random variable $\mathbf{B}^{(2)} \circ \phi(x)$ in $\mathbb{M}^2$ is indeed the D'Alembertian for the massless scalar field. We have already seen that the expectation value of the number of elements that lie in $L_0^-(x)$, $L_1^-(x)$ and $L_2^-(x)$ are, respectively,

$$\begin{aligned} \langle \mathbf{N}_0(x) \rangle &= \int_{y \in J^-(x)} dV_y e^{-\chi}, \\ \langle \mathbf{N}_1(x) \rangle &= \int_{y \in J^-(x)} dV_y \chi e^{-\chi}, \\ \langle \mathbf{N}_2(x) \rangle &= \int_{y \in J^-(x)} dV_y \frac{1}{2}\chi^2 e^{-\chi}, \end{aligned} \tag{3.73}$$

where $\chi \equiv \rho \mathrm{vol}(x, y)$. In the D'Alembertian, each of these comes with the weight $\phi(y)$ and hence

$$\langle \mathbf{B}^{(2)} \circ \phi(x) \rangle \equiv \int dV_y B(x-y)\phi(y) = 4\rho\Big(-\frac{1}{2}\phi(x) + \rho \int dV_y e^{-\chi}\Big(1 - 2\chi + \frac{1}{2}\chi^2\Big)\phi(y)\Big) \tag{3.74}$$

In the lightcone coordinates (u, v) with x at $(0, 0)$, $\chi = \rho uv$. We will impose an IR cut-off R in the integrals before taking the $\rho \to \infty$ limit, i.e., $u, v \in (-R, 0)$:

$$\begin{aligned} \langle \mathbf{B}^{(2)} \circ \phi(x) \rangle &= -2\rho\phi(x) + 4\rho^2 \int_{-R}^{0} du \int_{-R}^{0} dv e^{-\rho uv}\Big(1 - 2\rho uv + \frac{1}{2}\rho^2 u^2 v^2\Big)\phi(u, v) \\ &= -2\rho\phi(x) + 4\rho^2 \hat{O} \circ I_R, \end{aligned} \tag{3.75}$$

where $I_R \equiv \int_{-R}^{0} du \int_{-R}^{0} dv e^{-\rho uv}\phi(u, v)$ and the operator $\hat{O} \equiv \left(1 + 2\rho\frac{\partial}{\partial\rho} + \frac{1}{2}\rho^2\frac{\partial^2}{\partial\rho^2}\right)$.

Expanding $\phi(u, v)$ about $(0, 0)$,

$$\begin{aligned} \phi(u, v) =& \phi(x) + u\frac{\partial}{\partial u}\phi(x) + v\frac{\partial}{\partial v}\phi(x) + uv\frac{\partial^2}{\partial u \partial v}\phi(x) \\ &+ \frac{1}{2}u^2\frac{\partial^2}{\partial u^2}\phi(x) + \frac{1}{2}v^2\frac{\partial^2}{\partial v^2}\phi(x) + \ldots, \end{aligned} \tag{3.76}$$

allows the integral I_R in Eq. (3.75) to be written as a sum over a series of integrals

$$I_R = \sum_{r=0}^{\infty}\sum_{m=0}^{r} \frac{1}{m!(m-j)!}\phi_{(m,r-m)}(0) I_R(m, r-m),$$

$$I_R(m,n) \equiv \int_{-R}^{0} du \int_{-R}^{0} dv u^m v^n e^{-\rho uv}, \tag{3.77}$$

where $\phi_{(m,r-m)}(0) \equiv \frac{\partial^r}{\partial u^m \partial v^{r-m}}\phi(u,v)|_{(0,0)}$. Simplifying,

$$I_R(m,n) = (-1)^n \frac{\partial^n}{\partial \rho^n}\int_{-R}^{0} du\, u^{m-n-1}\left(\frac{e^{\rho R u}-1}{\rho}\right) = (-1)^n \frac{\partial^n}{\partial \rho^n}\widetilde{I}_R(m,n), \tag{3.78}$$

where

$$\widetilde{I}_R(m,n) = \begin{cases} \frac{1}{\rho} Ein(\rho R^2) & \text{for } m = n \\ \frac{1}{\rho R^{m-n}}\frac{\partial^{m-n}}{\partial \rho^{m-n}} Ein(\rho R^2) + (-)^{m-n}\frac{R^{m-n}}{\rho(m-n)} & \text{for } m > n, n \geq 0 \end{cases}$$

and

$$Ein(x) = \int_{-x}^{0} \frac{e^t - 1}{t} dt = \sum_{i=1}^{\infty} \frac{(-1)^{i+1} x^i}{i\, i!}. \tag{3.79}$$

Our interest is in the large ρ contribution to Eq. (3.75) which we find by evaluating the term by term contribution to

$$4\rho^2 \hat{O} \circ I_R = \sum_{r=0}^{\infty}\sum_{m=0}^{r} \frac{1}{j!(m-j)!}\phi_{(m,r-m)}(0)\left(4\rho^2 \hat{O} \circ (-1)^n I_R(m, r-m)\right). \tag{3.80}$$

Note that

$$\frac{\partial^n}{\partial \rho^n} Ein(\rho R^2) = \frac{(-1)^{n-1}(n-1)!}{\rho^n} + O(e^{-\rho R^2}) \tag{3.81}$$

where $O(e^{-\rho R^2})$ denotes an overall exponential decay. The following identities will also be useful:

$$\hat{O} \circ (AB) = B(\hat{O} \circ A) + 2\rho\left(1 + \frac{1}{4}\rho\frac{\partial}{\partial \rho}\right)\left(A\frac{\partial}{\partial \rho} B\right) \tag{3.82}$$

$$\hat{O} \circ \left(\frac{1}{\rho^m}\right) = \frac{1}{2\rho^m}\left(m^2 - m + 2\right) \Rightarrow \hat{O} \circ \left(\frac{1}{\rho}\right) = \hat{O} \circ \left(\frac{1}{\rho^2}\right) = 0, \tag{3.83}$$

$$\hat{O} \circ \left(\frac{1}{\rho^m} Ein(\rho R^2)\right) = \frac{1}{2\rho^m}\left((m^2 - m + 2) Ein(\rho R^2) + (4 - (m+1))\right) + O(e^{-\rho R^2}). \tag{3.84}$$

The $m = n = 0$ contribution to Eq. (3.77) is thus

$$4\rho^2 \hat{O} \circ I_R(0,0)\phi(0) = 2\rho\phi(0,0) - O(e^{-\rho R^2}), \tag{3.85}$$

the leading term canceling the $-2\rho\phi$ in Eqn (3.75). Next, the contribution from $I_R(1,0) = I_R(0,1) = \frac{1}{\rho^2} - \frac{R}{\rho} + O(e^{-\rho R^2})$ decays exponentially because of Eq. (3.83). The most significant contribution is from $I_R(1,1) = -\partial_\rho(\rho^{-1} Ein(\rho R^2))$,

$$4\rho^2\phi_{(1,1)}(0)\hat{O} \circ I_R(1,1) = -2\phi_{(1,1)}(0)(1 + O(e^{-\rho R^2})), \tag{3.86}$$

which (pause for effect) *is* the continuum D'Alembertian with correction terms! The leading order behaviour of the other terms is

$$I_R(m,n) = \begin{cases} \frac{a}{\rho^{m+1}} + \frac{b}{\rho^{m+1}} Ein(\rho R^2) + O(e^{-\rho R^2}) & \text{for } m = n > 0 \\ \frac{a}{\rho^{m+1}} + \frac{b}{\rho^{n+1}} + O(e^{-\rho R^2}) m > n, n \geq 0 \end{cases}$$

Operating with $4\rho^2\hat{O}$ gives terms that are of sub-leading order, where we use the asymptotic form

$$Ein(x) \to \ln(x), \text{ as } x \to \infty. \tag{3.87}$$

Thus, for $d = 2$,

$$\lim_{\rho\to\infty} \langle \mathbf{B}^{(2)} \circ \phi(x) \rangle = \Box^{(2)}\phi(x). \tag{3.88}$$

What emerges in this calculation is that the operator $\hat{O}$ must be chosen in such a way as to ensure that the divergent terms in $I_R(m,n)$ are killed.

The generalisation of $\hat{O}$ for $d = 4$ was found in [36] and for more general d in [37, 138]. The discrete D'Alembertian operator in dimension d is given by the expression [36, 37, 137, 138]

$$B^{(d)}\phi(x) = \rho^{\frac{2}{d}} \left(\alpha_d \phi(x) + \beta_d \sum_{i=1}^{n_d} C_i^{(d)} \sum_{y \in L_i} \phi(y) \right). \tag{3.89}$$

It limits to the continuum D'Alembertian

$$\lim_{\rho\to\infty} \langle \mathbf{B}^{(d)} \circ \phi(x) \rangle = \Box^{(d)}\phi(x), \tag{3.90}$$

when

$$\alpha_d = \begin{cases} -\frac{1}{\Gamma\left(1+\frac{2}{d}\right)} c_d^{2/d} & d \text{ odd} \\ -\frac{2}{\Gamma\left(1+\frac{2}{d}\right)} c_d^{2/d} & d \text{ even,} \end{cases} \tag{3.91}$$

$$\beta_d = \begin{cases} \dfrac{d+1}{2^{d-1}\Gamma\left(1+\frac{2}{d}\right)} c_d^{2/d} & d \text{ odd} \\ \dfrac{\Gamma\left(\frac{d}{2}+2\right)\Gamma\left(\frac{d}{2}+1\right)}{\Gamma\left(\frac{2}{d}\right)\Gamma(d)} c_d^{2/d} & d \text{ even} \end{cases} \tag{3.92}$$

and

$$n_d = \begin{cases} \frac{d}{2}+\frac{3}{2} & d \text{ odd} \\ \frac{d}{2}+2 & d \text{ even} \end{cases} \tag{3.93}$$

where $c_d = 2^{1-\frac{d}{2}} S_{d-2}/(d(d-1))$ (recall that S_d is the volume of the unit d-sphere), and

$$C_i^{(d)} = {}_{q+1}F_q(\{a_1, \ldots, a_q, i-1\}, \{b_1, \ldots b_q\}|1), \tag{3.94}$$

with $q = \frac{d+1}{2}$, $a_i = \frac{d+2i}{d}$, $b_i = \frac{2i}{d}$ for d odd and $q = \frac{d}{2}$, $a_i = \frac{d+2i+2}{d}$, and $b_i = \frac{2i}{d}$ for d even.

While this result is, of course, important in its own right, it leads to one of the most significant results in CST when generalised to a Riemann normal neighbourhood (RNN) in a generic spacetime. It was shown by Benincasa and Dowker in $d = 2, 4$ [36] and generalised to all dimensions d by Dowker and Glaser [37] that

$$\lim_{\rho\to\infty} \langle \mathbf{B}^{(d)} \circ \phi(x) \rangle = \Box^{(d)}\phi(x) - \frac{1}{2}R(x)\phi(x). \tag{3.95}$$

where $R(x)$ is the Ricci scalar associated with the RNN. For $\phi =$ constant this gives an expression for $R(x)$ in terms of $\mathbf{B}^{(d)} \circ \phi(x)$. Thus, the discrete Ricci curvature at an element $e \in C$ can be defined as

$$R^{(d)}(e) = -2\rho^{\frac{2}{d}}\left(\alpha_d + \beta_d \sum_{i=0}^{n_d} C_i^{(d)} N_i(e)\right), \tag{3.96}$$

where $N_k(e) = |L_k(e)|$. Summing over the n elements of a finite element causal set gives the discrete **Benincasa-Dowker-Glaser(BDG) action**

$$\frac{1}{\hbar} S_{BDG}^{(d)}(C) = -\alpha_d \left(\frac{\ell}{\ell_p}\right)^{d-2} \left(n + \frac{\beta_d}{\alpha_d} \sum_{i=0}^{n_d} C_i^{(d)} N_i\right), \tag{3.97}$$

where $\ell = \rho^{-\frac{1}{d}}$ and N_k is the total number of k-intervals in C. It was shown in [36,37] that

$$\lim_{\rho\to\infty} \langle S_{BDG}^{(d)}(C) \rangle = S_{EH}^{(d)}(g), \tag{3.98}$$

where $\ell = \rho^{-d}$, upto (as yet unknown) boundary terms (and where $S_{EH}^{(d)}(g)$ is the Einstein-Hilbert action.)

The contribution from the neighbourhood of $\partial J^-(x)$ away from the origin, can however be non-trivial. Using Fermi Normal Coordinates it was shown in [139] that this contribution is negligible in an approximately flat region of a four dimensional spacetime. Though simulations for Poisson sprinklings into different spacetimes support this more generally, an analytic extension of this calculation to an arbitrary spacetimes has been difficult to obtain.

Exercise 3.16 Calculate and contrast the BDG action with the action defined using Eq. (3.31) for the $n = 16$ KR order shown in Fig. A.7 and the $n = 16$ element causal set in Fig. A.8.

The construction of the CST action is an extremely significant development in CST, making it possible to study continuum inspired causal set dynamics, as we will see in Chap. 5.

Of course, as in all our discussions, we must worry about fluctuations. Since the discreteness scale appears explicitly in the discrete action, via the number of links N_0, these fluctuations can be expected to be large. It was shown in [134] that for $C \sim_\rho \mathbb{M}^2$, the fluctuations can be suppressed significantly if the contribution is "smeared" out over several layers $L_i(e)$ for each $e \in C$, determined by an intermediate **mesoscale**. Thus one gets a family of D'Alembertian operators

$$B_\epsilon^{(d)}\phi(e) = \epsilon^{\frac{2}{d}}\rho^{\frac{2}{d}}\left(\alpha_d\phi(e) + \beta_d\epsilon \sum_{e' \prec e} f_d(|(e,e')|,\epsilon)\phi(e')\right), \tag{3.99}$$

where $|(e, e')|$ is the cardinality of the interval (e, e'), the function

$$f_d(j,\epsilon) \equiv (1-\epsilon)^j \sum_i^{n_d} C_i^{(d)} \binom{n}{i-1}\left(\frac{\epsilon}{1-\epsilon}\right)^{i-1}, \tag{3.100}$$

and the mesoscale $\rho_\epsilon \equiv \epsilon\rho$, $\epsilon \in (0, 1]$. The function $f_d(n(e, e'), \epsilon)$ does the job of smearing over a clutch of layers as shown in Fig. 3.13. The first set of layers comes with a positive sign, the next with a negative sign and so on. Again, this was shown to converge to the continuum D'Alembertian [36,37]. Moreover, it also gives rise to a 1-parameter family of actions

$$S_\epsilon^{(d)}(C) = -\alpha_d\epsilon^{\frac{2}{d}}\rho^{\frac{2}{d}}\left(n + \frac{\beta_d}{\alpha_d}\epsilon \sum_{e \in C}\sum_{e' \prec e} f_d(n(e,e'),\epsilon)\right), \tag{3.101}$$

which again can be shown to limit to the Einstein-Hilbert action [36,37].

In the calculation above, we have taken $\phi(x)$ to be of compact support, which means that there are no boundary terms. As we have seen in Sect. 3.4 the discrete spatial GHY boundary terms can be constructed, but there is no immediate connection to the action we have constructed above. It is therefore of interest to ask if the above action contains *any* boundary terms at all. In [118] this question was considered for

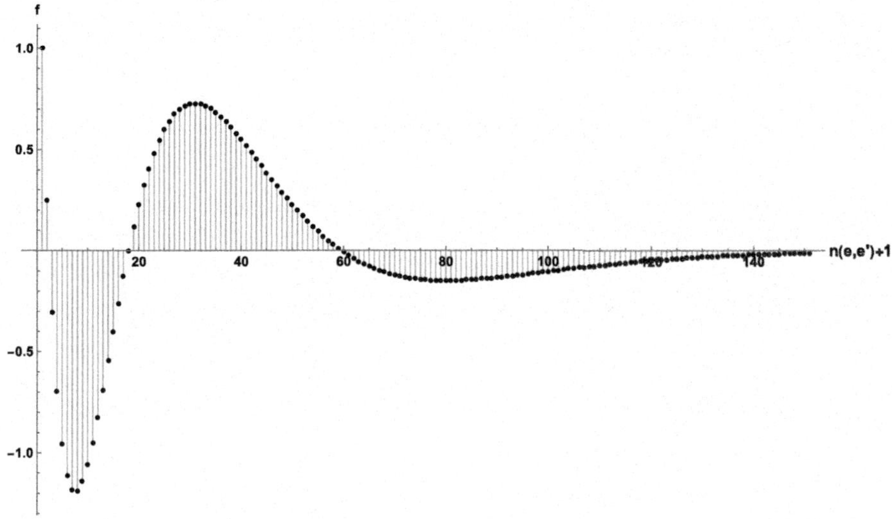

Fig. 3.13 The function $f(n, 0.05)$. There are 4 regions of alternating sign corresponding to 4 "smeared out" layers

a causal diamond in $\mathbb{M}^d$. The "bulk" action of course vanishes, but there is a residual boundary contribution.

We will check this for $C \in \mathfrak{C}_\rho(\mathbb{D}^d)$. The boundary $\partial(\mathbb{D}^d)$ has the null components $\mathcal{N}^{\pm}$ as well as the spatial codimension 2 joint $\mathcal{J}^{(d-2)} \sim \mathbb{S}^{d-2}$ which is a round $(d-2)$ sphere.

The expectation value of the BDG action (for $\epsilon = 1$)

$$\langle S^{(d)}(C)\rangle = -\alpha_d \left(\frac{\ell}{\ell_p}\right)^{d-2} \left(n + \frac{\beta_d}{\alpha_d} \sum_{i=0}^{n_d} C_i^{(d)} \langle N_i \rangle\right), \tag{3.102}$$

can be evaluated analytically using Eq. (3.68), which is equivalent to the more messy but useful expansion in n

$$\left\langle N_i^{(d)} \right\rangle = \frac{\Gamma(d)^2 n^{i+2}}{\Gamma(i)} \sum_{k=0}^{\infty} \frac{(-n)^k \, \Gamma(k+i+1)\, \Gamma\left(\frac{d(k+i)}{2}+1\right) \Gamma\left(\frac{d(k+i+1)}{2}+1\right)}{\Gamma(k+i+3)\, \Gamma(k+1)\, \Gamma\left(\frac{d(k+i)}{2}+d\right) \Gamma\left(\frac{d(k+i+1)}{2}+d\right)}. \tag{3.103}$$

For $d = 2$,

$$\frac{1}{\hbar}\left\langle S_{BDG}^{(2)} \right\rangle = 2\left(n - 2\left\langle N_1^{(2)}\right\rangle + 4\left\langle N_2^{(2)}\right\rangle - 2\left\langle N_3^{(2)}\right\rangle\right), \tag{3.104}$$

is straightforward to evaluate [118, 140] and yields

$$\frac{1}{\hbar}\left\langle S_{BDG}^{(2)} \right\rangle = 2(1 - e^{-n}). \tag{3.105}$$

In anticipation of the results for higher d we note that the volume of the zero sphere at the joint, $\mathrm{vol}(\mathcal{J}^{(0)}) = S_0 = 2$ and hence

$$\lim_{n\to\infty} \frac{1}{\hbar}\left\langle S^{(2)}_{BDG}\right\rangle = \mathrm{vol}(\mathcal{J}^{(0)}). \tag{3.106}$$

In other words, the causal set action of the causal diamond in $\mathbb{M}^2$ is *not* zero. If one were to stop at $d = 2$, multiple interpretations are possible, including the suggestion that S_{BDG} contains a topological term [141]. That this is in fact a boundary term is confirmed by doing the slightly more involved calculation in $d = 3$,

$$\frac{1}{\hbar}\left\langle S^{(3)}_{BDG}\right\rangle = -\alpha_3\left(\frac{l}{l_p}\right)\left(n - \left\langle N^{(3)}_1\right\rangle + \frac{27}{8}\left\langle N^{(3)}_2\right\rangle - \frac{9}{4}\left\langle N^{(3)}_3\right\rangle\right). \tag{3.107}$$

which is a power series expansion in n with coefficients

$$-\alpha_3\left(\frac{l}{l_p}\right) \times \frac{(-1)^{i+1}}{i!}\frac{8}{(3i+1)(3i-1)}, \tag{3.108}$$

$i = 1, \ldots \infty$. Again, rearranging indices gives the closed form expression

$$\frac{1}{\hbar}\left\langle S^{(3)}_{BDG}\right\rangle = -8\alpha_3\left(\frac{l}{l_p}\right)\left(-1 + \frac{1}{6n^{\frac{1}{3}}}\gamma(\frac{1}{3}, n) - \frac{n^{\frac{1}{3}}}{6}\gamma(-\frac{1}{3}, n)\right) \tag{3.109}$$

where $\gamma(s, x) \equiv \int_0^x t^{s-1}e^{-t}dt$ is a lower incomplete Gamma function. Taking the large n limit we find that

$$\lim_{N\to\infty} \frac{1}{\hbar}\langle S^{(3)}_{BDG}\rangle = -4\frac{1}{l_p}\alpha_3 l n^{1/3}\Gamma(2/3) = \frac{1}{l_p}\mathrm{vol}(\mathcal{J}^{(1)}). \tag{3.110}$$

In other words, not only does the action S_{BDG} contain a boundary term, but also, it is precisely the area of the joint $\mathcal{J}^{(1)}$ in $\mathbb{D}^3$, with no contribution from the null boundaries $\mathcal{N}^{\pm}$!

For general d it is easiest to start with the ratio

$$\frac{l_p^{d-2}\left\langle S^{(d)}_{BDG}\right\rangle}{\hbar\, \mathrm{vol}(\mathcal{J}^{(d-2)})} = \frac{\varepsilon_d}{d(d-1)\Gamma\left(1+\frac{2}{d}\right)n^{\frac{d-2}{d}}}\left(n + \frac{\beta_d}{\alpha_d}\sum_{i=1}^{n_d} C^{(d)}_i\left\langle N_i\right\rangle\right), \tag{3.111}$$

where $\varepsilon_d = 1$ for d odd and 2 for d even. A somewhat lengthy calculation [118] demonstrates (after a rearrangement and redefinition of indices) that

$$\sum_{i=1}^{n_d} C^{(d)}_i\left\langle N_i\right\rangle = \sum_{j=1}^{\infty} A^{(d)}_j n^{j+1}. \tag{3.112}$$

where

$$A_j^{(d)} = \Gamma(d)^2 \frac{(-1)^j}{(j+1)!} \frac{\Gamma\left(\frac{d}{2}(j-1)+1\right)\Gamma\left(\frac{d}{2}j+1\right)}{\Gamma\left(\frac{d}{2}(j-1)+d\right)\Gamma\left(\frac{d}{2}j+d\right)} \sum_{i=1}^{D+2} (-1)^i \binom{j-1}{i-1} C_i^{(d)}, \tag{3.113}$$

and $d = 2D$ for d even and $d = 2D + 1$ for d odd. Evaluating this sum is non-trivial, even using symbolic computational tools like those available in *Mathematica*. Using various intermediate simplifications one can nevertheless show for $d = 2, \ldots, 20$ (and possibly higher) that

$$A_j^{(d)} = \frac{\Gamma(d)^2 (-1)^{j+1}}{\Gamma\left(\frac{d}{2}(j+1)\right)\Gamma\left(\frac{d}{2}(2+j)\right)\Gamma(2+j)} \gamma_j^{(d)}, \tag{3.114}$$

where

$$\gamma_j^{(d)} = \begin{cases} \dfrac{\sqrt{\pi}}{2^{1+dj}} \dfrac{\Gamma(2+dj)}{\Gamma\left(\frac{d-1}{2}\right)} & d \text{ odd} \\ \dfrac{\Gamma\left(1+\frac{d}{2}j\right)\Gamma\left(2+\frac{d}{2}j\right)}{\Gamma\left(\frac{d}{2}\right)} & d \text{ even.} \end{cases} \tag{3.115}$$

Inserting (3.114) into (3.111) gives, for $d = 2, \ldots, 20$

$$\lim_{N\to\infty} \frac{1}{\hbar}\left\langle S_{BDG}^{d} \right\rangle = \frac{1}{l_p^{d-2}} \mathrm{vol}(\mathcal{J}^{(d-2)}). \tag{3.116}$$

Thus, the discrete BDG action for $\mathfrak{C}_\rho(\mathbb{D}^d)$ contains a special "joint" boundary term, but with no contribution from the null boundaries.

This verifies a conjecture due to Benincasa and Dowker that Eq. (3.116) holds more generally for causal diamonds in all spacetimes. A generalisation to causal intervals in a RNN in an arbitrary spacetime [142,143] provides further verification of the conjecture, though the techniques used are somewhat different.

3.6 Summary

There are several other order invariants which have a continuum geometric interpretation like horizon molecules, null geodesics, proper distance, timelike boundaries as well as the spectral dimension. It is not possible to cover all these results in the present text and we point the interested reader to this literature [129,130,144–154].

In this chapter we considered explicit examples of order invariants in $\mathfrak{C}_\rho(M, g)$ and showed explicitly (sometimes for specific (M, g)) that they correspond to geometric and topological invariants of (M, g). These order invariants were built from a set of three very simple order theoretic structures – the chains, the antichains and the intervals. Despite their simplicity, they were seen to contain a rich amount of continuum information.

Although we do not yet have a proof of the Hauptvermutung, these examples give us the confidence that $\mathfrak{C}_\rho(M, g)$ contains physically relevant geometric information. Hence relevant geometric information is not "lost" by assuming discreteness! These order invariants are moreover an important class of discrete covariant observables for the quantum dynamics of CST, and can be used, for example, to distinguish between the continuumlike and non-continuumlike phases of the theory (see Chap. 5). Thus, understanding and characterising the kinematics of CST gives us important tools for studying the full quantum dynamics of causal sets.

The kinematical discretisation of the continuum which we have explored in this chapter can be thought of as a "first quantisation", in which the classical spacetime structure is replaced by an atomistic, probabilistic ensemble. Understanding it in more detail is essential to the construction of its "second quantisation", i.e., a fully non-perturbative quantisation of causal sets. As we will see in Chap. 4 the first quantised framework also provides a covariant regularisation of Quantum Field Theory and can have far reaching observable consequences, many of which are yet to be explored.

Quantum Scalar Fields on Causal Sets

4

Abstract

On the way to constructing a full dynamics for CST, lies the phenomenologically interesting regime of quantum field theory (QFT) on an ensemble of continuumlike causal sets. The kinematical discreteness of causal sets provides a natural covariant spacetime cut-off, which not only regulates, but can also modify the nature of quantum fields in the deep UV. QFT on causal sets is formulated using the Sorkin-Johnston framework which uses the causal structure. Starting with the advanced and retarded Green's functions and the spacetime commutator or Peierls bracket, the Sorkin-Johnston (SJ) Wightmann function of a free Gaussian scalar field can be defined for continuumlike causal sets. This in turn can be used to construct Sorkin's spacetime entanglement entropy (SSEE) for the free scalar field between sub-regions of a causal set.

4.1 Introduction

The tensions that arise from quantising matter but *not* gravity have perhaps provided some of the most compelling reasons for seeking an underlying quantum theory of gravity. The hybrid Einstein field equations

$$G_{ab} = 8\pi \langle T_{ab} \rangle \tag{4.1}$$

involve the classical gravitational field on one side and the vacuum expectation value of the quantum matter stress energy tensor on the other. This dichotomy is rife with interpretational issues, if treated as fundamental. However, one can make physical sense of Eq. (4.1) in certain energy regimes of phenomenological interest, by regarding the classical gravitational field as a collective excitation of more fundamental quantum gravity degrees of freedom.

The study of quantum fields on a fixed curved spacetime therefore provides an interesting playground for quantum gravity, challenging many aspects of quantum field theory on Minkowski spacetime. The puzzles of particle production at horizons,

S. Surya, *The Causal Set Approach to Quantum Gravity*, Lecture Notes in Physics 1036,
https://doi.org/10.1007/978-3-031-84420-1_4

for example, provide possible clues to quantum gravity and are a subject of intense investigation in several approaches to quantum gravity.

In this chapter we study quantum scalar field theory on a kinematic ensemble of causal sets $\mathfrak{C}_\rho(M, g)$ for a fixed continuum spacetime (M, g). This kinematic regime of CST, where spacetime is replaced by the underlying ensemble of causal sets but is not otherwise dynamically quantised, provides an excellent in-between ground for formulating these questions. An important feature of CST is that it provides a covariant spacetime "UV" cut-off, which holds the promise of resolving the divergences of QFT.

This use of discreteness without dynamical quantisation may sound counter-intuitive to a reader who has skipped Chap. 3, since discreteness makes some assumptions about quantum gravity. However, at this kinematical level, this is not conceptually very different from the study of quantum field theory on $4+d$ dimensional Kaluza-Klein spacetimes, where d is the dimension of a compactified space. In both cases, one is interested in "beyond standard model" physics in which quantum gravity plays an effective role in changing the background structure. This effective role can be realised either by increasing the spacetime dimension, or in the case of causal sets, by the discretisation of the causal structure.

An important challenge that such an endeavour faces is that typically, QFT is formulated using equal time commutation relations. We know that there is no meaningful analogue of Cauchy hypersurfaces as carriers of initial data in a continuumlike causal set. Hence we need a more spacetime formulation which, moreover, relies on the causal structure of the spacetime. This is the Sorkin-Johnston prescription for free (Gaussian) scalar fields, which was first developed by Johnston for QFT on causal sets [38,39,155] and later shown by Sorkin to give an observer independent preferred vacuum state [40,156]. To begin with, the equal time commutation relations are replaced by the spacetime commutator, or Peierls bracket using the Pauli-Jordan function which is given by the difference in the retarded and advanced Green's functions.

The first task is therefore to identify the causal set Green's functions for a given $\mathfrak{C}_\rho(M, g)$. In Chap. 3 we noticed that for $\mathfrak{C}_\rho(\mathbb{M}^2)$, the retarded scalar Green's function is given by the causal matrix Eq. (3.33). In Sect. 4.2 we will show constructions of the advanced and retarded scalar Green's functions on causal sets that are approximated by $\mathbb{M}^4$, and $d = 2, 4$-dimensional de Sitter dS_d and anti-de Sitter adS_d spacetimes, as well as more general spacetime Riemann normal neighbourhoods in $d = 2, 4$ [155,157,158].

In Sect. 4.3 we describe Johnston's quantisation of the free scalar field using the covariant Peierls bracket formulation [38–40]. The associated Pauli-Jordan (PJ) function $i\Delta(x, x')$, which is i times the difference between the retarded and advanced Green's functions, is elevated to an integral operator $i\widehat{\Delta}$. The eigenmodes of this operator provide an observer independent decomposition into positive and negative modes of the quantum field. The Sorkin-Johnston (SJ) vacuum, or Wightmann function, defined as the positive part of $i\widehat{\Delta}$, depends on the global features of the spacetime, but is observer independent and thus "distinguished" [156]. This is a continuum prescription [156,159–161] which has an explicit causal set analogue [38,41,158].

In Sect. 4.4 we define Sorkin's spacetime entanglement entropy (SSEE) for causally complementary regions in a finite spacetime region [42, 162]. This construction carries over naturally to finite causal sets [43,44,158,163–165]. We discuss how the SSEE for a class of continuumlike causal sets satisfies *both* a volume relation as well as an area relation, depending on the scale. While an the area is obtained via a double truncation of the spectrum of $i\widehat{\Delta}$, as first shown by Sorkin and Yazdi [43], the non-trivial UV behaviour leads to an unexpected volume law. We discuss the implications of this scale dependent entanglement entropy, and the possible role of nonlocality.

Recent work on interacting field theories is very promising and could lead to important phenomenological predictions from the theory [45–47]. However a pedagogical discussion on this is not possible in this book, and we refer the interested reader to the literature. Being perturbative, the analysis of the interacting fields in [47] also rests on the constructions in this current chapter.

4.2 Green's Functions on Causal Sets

The massless retarded free scalar Green's function $G_R(x, x')$ in $\mathbb{M}^d$ has a non-trivial dependence on d. This is reflected in the failure of Huygen's principle in odd dimensions $d \geq 3$ as well as in $d = 2$, where $G_R(x, x')$ has non-vanishing support inside the light cone. This is also the case for generic curved spacetimes in any dimension, due to the non-linearity of the D'Alembertian. It is therefore clear that the Green's function on a continuumlike causal set cannot have a single, dimension-independent, order theoretic definition. Instead, as in the continuum, one has to examine each case separately.

In this section we will construct causal scalar Green's functions on causal sets that are approximated by certain spacetimes whose Green's functions have a simple functional form. As in Chap. 3, the aim is to obtain an intrinsic, order theoretic characterisation of the Green's function in each case.

Consider a massive scalar field theory in a spacetime (M, g) with non-minimal coupling ξ. Its Green's function $G_{m,\xi}(x, x')$ satisfies

$$(\Box_g - m^2 - \xi R)G_{m,\xi}(x, x') = \frac{1}{\sqrt{-g(x)}}\delta(x - x'). \tag{4.2}$$

The $m \neq 0$ Green's function can be formally generated from $G_{0,\xi}(x, x')$ via the relation

$$\begin{aligned}
G_{m,\xi}(x, x') &= G_{m,\xi}(x, x') - m^2\, G_{0,\xi} * G_{0,\xi}(x, x') + m^4\, G_{0,\xi} * G_{0,\xi} * G_{0,\xi}(x, x') + \ldots \\
&= \sum_{k=0}^{\infty}(-m^2)^k \underbrace{G_{0,\xi} * G_{0,\xi} * \ldots G_{0,\xi}(x, x')}_{k+1} \\
&= G_{0,\xi}(x, x') - m^2 G_{0,\xi} * G_{m,\xi}(x, x')
\end{aligned} \tag{4.3}$$

where the convolution is defined as

$$(A * B)(x, x') \equiv \int d^d y \sqrt{-g(y)} A(x, y) B(y, x'). \tag{4.4}$$

Operating on Eq. (4.3) with the massless Klein-Gordon operator $(\Box_g - \xi R)$ and using the massless Green's equation, we recover the massive Green's equation Eq. (4.2). We can similarly generate the massless Green's function from the massive one. For the special case of constant Ricci curvature R, define the generalised mass $M^2 = m^2 - \xi R$. $G_M(x, x')$ can thus be generated from $G_0(x, x')$ or any other $G_{M'}(x, x')$.

This is a useful prescription for the causal set Green's function. Starting with the causal set massless Green's function $K_{0,\xi}(e, e')$, we can generate the massive Green's function $K_{m,\xi}(e, e')$ using the discrete analog of Eq. (4.3)

$$\begin{aligned} K_{m,\xi}(e, e') &= K_{0,\xi}(e, e') - \frac{m^2}{\rho}(K_{0,\xi} * K_{0,\xi})(e, e') + \frac{m^4}{\rho^2}(K_{0,\xi} * K_{0,\xi} * K_{0,\xi})(e, e') + \ldots \\ &= \sum_{k=0}^{\infty} \left(-\frac{m^2}{\rho}\right)^k \underbrace{\Big(K_{0,\xi} * K_{0,\xi} * \ldots K_{0,\xi}\Big)}_{k+1}(e, e') \end{aligned} \tag{4.5}$$

where now

$$(A * B)(e, e') \equiv \sum_{e \prec e'' \prec e'} A(e, e'') B(e'', e'). \tag{4.6}$$

The challenge for causal set theory is to find $K_{0,\xi}(e, e')$ for various $\mathfrak{C}_\rho(M, g)$. As also in the continuum, obtaining the Green's functions for an arbitrary spacetime (M, g) is non-trivial. We currently know the form of $K_{0,\xi}(e, e')$ for only a handful of cases in $d = 2, 4$: $\mathbb{M}^2$, $\mathbb{M}^4$, dS_2, dS_4, adS_2, adS_4, the RNN in $d = 2$ and special cases of the RNN in $d = 4$. In all the cases we consider here, $K_{0,\xi}(e, e')$ takes the same form as in $\mathbb{M}^d$. It is therefore important to study the case of $\mathfrak{C}_\rho(\mathbb{M}^d)$ first.

4.2.1 Discrete Green's Functions on $\mathbb{M}^d$

We have already discussed the causal set analog of the $d = 2$ massless retarded Green's function in Chap. 3, but repeat it here for completeness. The continuum Green's function is

$$G_0^{(2)}(x, x') = \frac{1}{2}\theta(x'_0 - x_0)\theta(\tau^2(x, x')), \tag{4.7}$$

which has uniform support inside the past light cone of $\frac{1}{2}$. For any causal set C, the causal matrix is defined as

$$M_C(e, e') \equiv \begin{cases} 1 \text{ if } e' \prec e \\ 0 \text{ otherwise} \end{cases} \tag{4.8}$$

The entries $M_C(e, e')$ corresponds to the existence of a 2-chain c_2 or, simply put, a relation, from e to e'. Its expectation value over $\mathfrak{C}_\rho(\mathbb{M}^2)$ is therefore

$$\langle \mathbf{M}_C(x, x') \rangle = \theta(x_0' - x_0)\theta(\tau^2(x, x')), \tag{4.9}$$

and hence a natural choice is

$$K_0^{(2)}(e, e') \equiv \frac{1}{2} M_C(e, e'), \Rightarrow \langle \mathbf{K}_0^{(2)}(x, x') \rangle = G_0^{(2)}(x, x'). \tag{4.10}$$

The massive retarded causal set Green's function is generated from $K_0^{(2)}(x, x')$ by

$$\begin{aligned} K_m^{(2)}(e, e') &= \sum_{k=0}^{\infty} \left(-\frac{m^2}{\rho}\right)^k \underbrace{\left(K_0^{(2)} * K_0^{(2)} * \dots K_0^{(2)}\right)}_{k+1}(e, e') \\ &= \sum_{k=0}^{\infty} \left(-\frac{m^2}{\rho}\right)^k \left(\frac{1}{2}\right)^{k+1} \underbrace{M_C * M_C * \dots M_C}_{k+1}(e, e') \\ &= \sum_{k=0}^{\infty} \left(-\frac{m^2}{\rho}\right)^k \left(\frac{1}{2}\right)^{k+1} C_{k+2}(e, e'), \end{aligned} \tag{4.11}$$

where $C_{k+2}(e, e')$ denotes the number of $(k+2)$-chains from e to e'. The sum terminates at the length of the longest chain from e and e'. Taking the average over all Poisson sprinklings

$$\langle \mathbf{K}_m^{(2)}(x, x') \rangle = \sum_{k=0}^{\infty} \left(-\frac{m^2}{\rho}\right)^k \left(\frac{1}{2}\right)^{k+1} \langle \mathbf{C}_{k+2}(x, x') \rangle, \tag{4.12}$$

where k goes all the way to infinity to allow for all possible sprinklings. This can be explicitly evaluated using Eq. (3.18), to give

$$\begin{aligned} \langle K_m^{(2)}(x, x') \rangle &= \theta(x_0' - x_0)\theta(\tau^2(x, x')) \sum_{k=0}^{\infty} (-1)^k \frac{1}{2} \left(\frac{m\tau}{2}\right)^{2k} \frac{1}{(k!)^2} \\ &= \frac{1}{2}\theta(x_0' - x_0)\theta(\tau^2(x, x')) J_0(m\tau) = G_m^{(2)}(x, x'), \end{aligned} \tag{4.13}$$

where $J_\alpha(x)$ is the Bessel function. This relationship holds for any ρ, not just the continuum limit $\rho \to \infty$. This is a special feature of $d = 2$, and as we will see, is not true more generally.

In $\mathbb{M}^4$

$$G_0^{(4)}(x, x') = \frac{1}{2\pi}\theta(x_0 - x_0')\delta(\tau^2(x, x')), \tag{4.14}$$

which has support only on the past light cone of x. The GPPP or Poisson sprinkling does not realise sets of zero spacetime volume which are sets of measure zero. Thus, codimension 1 and higher regions of a spacetime are not probed by the sprinkling. This includes lightcones. A natural choice to approximate a lightlike relation is the link relation $\prec\!\!*$ which though not strictly null, "hugs" the light cone (see [166] for other candidates.) [155]. In $d = 4$, instead of the causal matrix, we therefore use the **link matrix**

$$L_C(e, e') \equiv \begin{cases} 1 \text{ if } e' \prec\!\!* \, e \\ 0 \text{ otherwise.} \end{cases} \tag{4.15}$$

The probability for a given pair x, x' in the sprinkling to be linked is $e^{-\rho \mathrm{vol}(x,x')}$, where $\mathrm{vol}(x, x') = \frac{\pi}{24}\tau^4(x, x')$ is the volume of an Alexandrov interval $I(x, x')$ in $\mathbb{M}^4$. Thus,

$$\langle \mathbf{L}_C(x, x')\rangle = \theta(x_0 - x_0')\theta(\tau^2(x, x'))e^{-\frac{\rho\pi}{24}\tau^4(x,x')}, \tag{4.16}$$

which, in the continuum limit simplifies to

$$\lim_{\rho\to\infty} \sqrt{\frac{\rho}{6}}\langle \mathbf{L}_C(x, x')\rangle = 2\,\theta(x_0 - x_0')\theta(\tau^2)\delta(\tau^2) = 2\pi G_0^{(4)}(x, x')\,. \tag{4.17}$$

This gives us a candidate for the massless causal set Green's function on $\mathbb{M}^4$ [155],

$$K_0^{(4)}(e, e') = \frac{1}{2\pi}\sqrt{\frac{\rho}{6}}L_C(e, e')\,. \tag{4.18}$$

Unlike the case of $\mathbb{M}^2$ which is exact for any sprinkling density ρ, in $\mathbb{M}^4$ we see that the causal set Green's function reproduces the continuum counterpart only in the continuum limit. This suggests that there are potentially interesting phenomenological effects of discreteness. At finite ρ, the exponential function $\exp(-\frac{\rho\pi}{24}\tau^4(x, x'))$ in Eq. (4.16) leaks into the light cone (but never outside!), which in the continuum could be interpreted as a small, but detectable mass or curvature.

Next, we generate the massive $K_m^{(4)}(x, x')$ from $K_0^{(4)}(x, x')$

$$\begin{aligned} K_m^{(4)}(e, e') &= \sum_{k=0}^{\infty}\left(-\frac{m^2}{\rho}\right)^k \underbrace{K_0^{(4)} * K_0^{(4)} * \ldots K_0^{(4)}}_{k+1}(e, e') \\ &= \sum_{k=0}^{\infty}\left(-\frac{m^2}{\rho}\right)^k \left(\frac{1}{2\pi}\sqrt{\frac{\rho}{6}}\right)^{k+1} L_{k+2}(e, e') \end{aligned} \tag{4.19}$$

where

$$L_{k+2}(e, e') = \underbrace{L_C * L_C * \ldots L_C}_{k+1}(e, e') \tag{4.20}$$

denotes the number of $(k+2)$-element linked-paths, $e_1 \prec\!\!* \ldots e_i \prec\!\!* e_{i+1} \ldots \prec\!\!* e_{k+2}$ from $e_1 = e$ to $e_{k+2} = e'$. The sum in Eq. (4.19) terminates at the longest linked-path from e to e'. Using the statistical independence of the Poisson sprinkling,

$$\langle \mathbf{L}_{k+2}(x, x') \rangle = \rho^k (\underbrace{\langle L_C \rangle * \ldots * \langle L_C \rangle}_{k+1})(x, x') \tag{4.21}$$

and hence

$$\lim_{\rho\to\infty} \langle \mathbf{K}_m^{(4)}(x, x') \rangle = \lim_{\rho\to\infty} \sum_{k=0}^{\infty} (-m^2)^k \left(\frac{1}{2\pi}\sqrt{\frac{\rho}{6}}\right)^{k+1} \underbrace{\langle \mathbf{L}_C \rangle * \ldots * \langle \mathbf{L}_C \rangle}_{k+1}(x, x') \tag{4.22}$$

$$= \sum_{k=0}^{\infty} (-m^2)^k \underbrace{G_0^{(4)} * G_0^{(4)} * \ldots G_0^{(4)}}_{k+1}(x, x') \tag{4.23}$$

$$= G_m^{(4)}(x, x') \,. \tag{4.24}$$

Note that our reasoning in this last step differs from the $d = 2$ case. This is because the evaluation of $\langle \mathbf{L}_k(x, x') \rangle$ which involves nested integrals of $\exp(-\rho V(x, x'))$ is quite non-trivial, even in the large ρ limit and there is no known closed form expression for this.

Exercise 4.1 Write down $K_0^{(2)}$ and $K_0^{(4)}$ for the causal set in Fig. 4.1 and generate the massive Green's functions $K_m^{(2)}$, $K_m^{(4)}$.

Exercise 4.2 Using the continuum Green's function

$$G_m^{(4)}(x, x') = \theta(x_0 - x_0')\left(\frac{1}{2\pi}\delta(\tau^2(x, x')) - \theta(\tau^2)\frac{m}{4\pi\tau} J_1(m\tau)\right) \tag{4.25}$$

guess the leading order expression for $\langle \mathbf{L}_{k+2}(x, x') \rangle$ using the series form expression for $J_1(x)$.

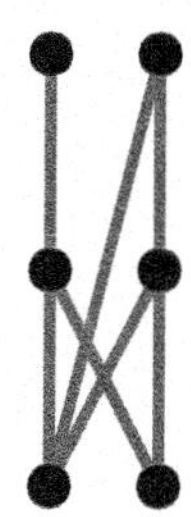

Fig. 4.1 Causal set for Exercise 4.1

Numerical simulations of $\langle \mathbf{L}_k(x, x') \rangle$ allow us to make a comparison between the discrete and continuum Green's functions at finite ρ, and to study the fluctuations about the mean in a systematic manner.

Exercise 4.3 In a simulation of an ensemble of causal sets $\mathfrak{C}_\rho(\mathbb{D}^2)$ one can construct $K_0^{(2)}(x, x')$ for any (fixed) pair in $x, x' \in I(p, q)$ and use it to generate $\langle \mathbf{K}_m^{(2)}(x, x') \rangle$. Comparisons with $G_m^{(2)}(x, x')$ however point to non-trivial fluctuations about the mean. What is the source of these fluctuations, and how do they change with m? *Hint: see [39].*

4.2.1.1 Hops and Stops

When Johston [155] first proposed the candidate Green's function in $\mathbb{M}^d$, he suggested a model using a sum-over-paths, and assigned weights to each **generalised causal trajectory** (GCT.) We describe this construction briefly here and some natural generalisations thereof.

Consider a particle hopping along on a causal set from e to e' along a GCT $\gamma_k(e, e') \equiv \{e_1, e_2, \ldots e_k\}$, where $e_i \prec e_{i+1}$, with $e_1 = e, e_k = e'$. Define $K(e, e')$ as the sum over all weighted trajectories

$$K(e, e') = \sum_{k=0}^{\infty} \Bigg(\sum_{\gamma_{k+2} \in \Gamma_{k+2}} w(\gamma_{k+2}(e, e')) \Bigg), \tag{4.26}$$

where Γ_{k+2} is the set of all $(k+2)$-element GCTs γ_{k+2} from e to e' and w is some choice of weight for each γ_k. Next, assign complex weights as follows: for every "hop" from $e_i \rightarrow e_{i+1}$ give a weight a, and for every "stop" at $e_i, i \neq 0, k+1$, give a weight b. This simplifies Eq. (4.26) to

$$K(e, e') = \sum_{k=0}^{\infty} a^{k+1} b^k T_{k+2}(e, e'), \tag{4.27}$$

where $T_{k+2}(e, e')$ is the number of $(k+2)$-trajectories from e to e'. In $\mathbb{M}^2$, for example, the $(k+2)$-trajectories are the $(k+2)$-chains, while in $\mathbb{M}^4$ they are the $(k+2)$-linked-paths. Once can then immediately read off the hop and stop weights by comparison with Eq. (4.11), (4.19) (Fig. 4.2):

$$\mathrm{d} = 2: \; a = \frac{1}{2}, \; b = -\frac{m^2}{\rho}, \qquad \mathrm{d} = 4: \; a = \frac{1}{2\pi}\sqrt{\frac{\rho}{6}}, \; b = -\frac{m^2}{\rho}. \tag{4.28}$$

In a more general hop and stop model, one can include an additional weight for each GCT

$$K(e, e') \equiv \sum_{k=0}^{\infty} a^{k+1} b^k \sum_{\gamma_{k+2} \in \Gamma_{k+2}} w_{k+2}(\gamma_{k+2}(e, e')). \tag{4.29}$$

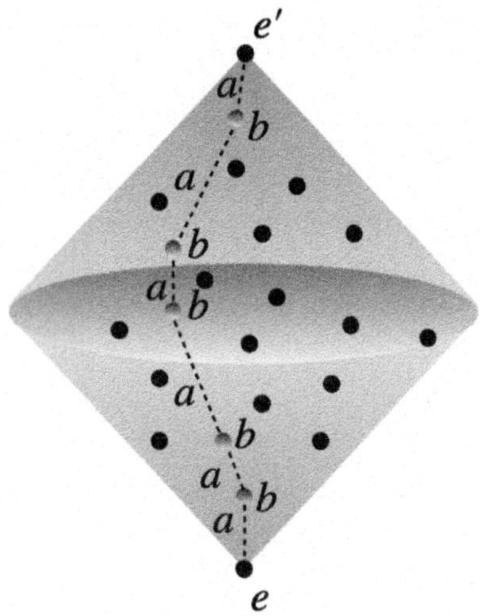

Fig. 4.2 A trajectory from e to e' in $C \sim_\rho \mathbb{M}^2$ with weights a for every hop, and weights b for every stop

If $w_{k+2}(\gamma_k(e, e'))$ is required to satisfy a Markovian factorisation condition

$$w_{k+2}(\gamma_{k+2}(e, e')) = w_2(\gamma_2(e, e_1))w_2(\gamma_2(e_1, e_2)) \ldots w_2(\gamma_2(e_k, e')), \tag{4.30}$$

then we can write

$$W_{k+2}(e, e') = \sum_{e_2} W_2(e, e_2) W_{k+1}(e_2, e') \tag{4.31}$$

where

$$W_{k+2}(e, e') \equiv \sum_{\gamma_{k+2} \in \Gamma_{k+2}} w_{k+2}(\gamma_{k+2}(e, e')), \tag{4.32}$$

with $W_2(e, e') \equiv w_2(\gamma_2(e, e'))$. For example, if the 2-element trajectory is a 2-chain and $w_2(\gamma_0(x, x')) = |I(x, x')|$, then using $\langle |I(x, x')| \rangle = \rho V(x, x')$, $\langle \mathbf{W}_2(x, x') \rangle = \theta(x_0' - x_0)\theta(\tau^2)\rho V(x, x')$. The factorisation property Eq. (4.31) and the independence of the sprinkling procedure implies that

$$\langle \mathbf{W}_{k+2}(x, x') \rangle = \rho \int d^d x_2 \sqrt{-g(x_2)} \langle \mathbf{W}_2(x, x_2) \rangle \langle \mathbf{W}_{k+1}(x_2, x') \rangle. \tag{4.33}$$

Defining $K_0(x, x') \equiv a W_2(x, x')$ we get the general form (for example, for the massive propagator)

$$\langle \mathbf{K}_b(x, x') \rangle = \langle \mathbf{K}_0(x, x') \rangle + b\rho \int d^d x_1 \sqrt{-g(x_1)} \langle \mathbf{K}_0(x, x_1) \rangle \, \langle \mathbf{K}_b(x_1, x') \rangle. \tag{4.34}$$

or equivalently

$$\langle \mathbf{K}_b(x, x') \rangle = \langle K_0(x, x') \rangle + b\rho(\langle \mathbf{K}_0 \rangle * \langle \mathbf{K}_0 \rangle)(x, x') + b^2\rho^2(\langle \mathbf{K}_0 \rangle * \langle \mathbf{K}_0 \rangle * \langle \mathbf{K}_0 \rangle)(x, x') \ldots \tag{4.35}$$

Thus, assuming the factorisation property Eq. (4.31) is satisfied, the identification of $K_0(x, x')$ in a spacetime boils down to choosing the appropriate kind of 2-trajectory $\gamma_2(x, x')$, as well as the weight $w_2(\gamma_0(x, x'))$.

Exercise 4.4 Verify that the expression above reduces to the massive K_m in $\mathbb{M}^2$ and $\mathbb{M}^4$.

As an illustration of this general framework, let us consider the Green's function in $\mathbb{M}^3$ where the massless retarded free scalar propagator is

$$G_0^{(3)}(x, x') = \theta(t' - t)\theta(\tau^2)\frac{1}{2\pi\tau(x, x')}. \tag{4.36}$$

As we saw in Chap. 3, the causal set counterpart of the proper time $\tau(x, x')$ in $\mathbb{M}^d$ is proportional to the length $l(x, x')$ of the longest chain from x to x' (see Eq. 3.8.) This suggests a causal set massless Green's function of the form

$$K_0^{(3)}(x, x') \equiv aH(x, x'), \tag{4.37}$$

where

$$H(x, x') \equiv \begin{cases} \dfrac{1}{l(x, x')} & \text{if } x \prec x' \\ 0 & \text{otherwise.} \end{cases} \tag{4.38}$$

Let us assume that the inverse relation holds, i.e.,

$$\lim_{\rho \to \infty} \langle \mathbf{H}(x, x')\rangle^{-1} \rho^{\frac{1}{3}} = \frac{1}{\kappa_3 \tau(x, x')}. \tag{4.39}$$

Comparing with Eq. (4.36) we find that $a = \rho^{1/3}\frac{\kappa_3}{2\pi} = (\frac{\rho\pi}{12})^{1/3}\frac{m_3}{2\pi}$. This is an example of a non-trivial weight $w_2(\gamma_2(x, x')) = \frac{1}{l(x,x')}$ associated with a 2-chain, i.e., the relation $x \prec x'$. Defining $H_2(e, e') \equiv H(e, e')$, the convolution

$$\langle \mathbf{H}\rangle * \langle \mathbf{H}\rangle = \rho \int d^3x_1 \langle \mathbf{H}(x, x_1)\rangle\langle \mathbf{H}(x_1, x')\rangle \equiv \langle \mathbf{H}_3(x, x')\rangle \tag{4.40}$$

where

$$H_3(x, x') = \sum_{x_1 \in C} H_2(x, x_1) H_2(x_1, x') = \sum_{x_1} \frac{1}{l(x, x_1)} M_C(x, x_1) \frac{1}{l(x_1, x')} M_C(x_1, x'). \tag{4.41}$$

$H_3(x, x')$ is the number of 3-chains from x to x' each weighted by the inverse of the maximum chain length from x to x_1 and x_1 to x' for all $x \prec x_1 \prec x'$. As in $d = 2$, the $(k + 2)$-trajectories are therefore chains, but the $H_{k+2}(x, x')$ are not obtained by merely counting the number of chains. Each $(k + 2)$-chain is additionally weighted by the inverse of the maximum chain length between each pair of joints, including the end points.

The massive causal set Green's function therefore takes the form

$$K_m^{(3)}(e,e') = \sum_{k=0}^{L} a^{k+1} b^k H_{k+2}(e,e') = \sum_{k=0}^{L} \left(\frac{m_3}{2\pi}\left(\frac{\pi\rho}{12}\right)^{1/3}\right)^{k+1} \left(-\frac{m^2}{\rho}\right)^k H_{k+2}(e,e'). \tag{4.42}$$

Verifying this ansatz analytically is non-trivial. However numerical simulations provide support for Eq. (4.39) [158].

Finally, massless propagators in the continuum are derivatives of either $\frac{1}{\tau}$ or $\delta(\tau^2)$ depending on whether d is odd or even. Since derivatives of $\delta(\tau^2)$ of any order can always be written as products of $\delta(\tau^2)$ and $\frac{1}{\tau}$, the knowledge of the causal set analogues of these two quantities with appropriate weights should in principle suffice to write down the discrete propagator. This is an example of an open, but well defined problem which should help generalise these constructions to higher dimensions.

4.2.2 Discrete Green's Functions in an RNN

The functional form of the scalar Green's functions are not known for generic spacetimes since they involve solving non-trivial second order PDEs. To study the UV properties of a field theory it suffices to look at the Riemann Normal Neighbourhood (RNN) O_p about an event p. Using the RNN coordinates, the spacetime metric can be expanded about p as

$$g_{ab}(x) = \eta_{ab} + \frac{1}{2!}\partial_c\partial_d\, g_{ab}(0)x^c x^d + O(x^3), \tag{4.43}$$

where p is taken to be at the origin $x = 0$ and $|Rx^2| \ll 1$. Bunch and Parker [167] calculated the Green's function in an RNN, starting from the Green's density

$$\mathbb{G}_{m,\xi}(x,x') \equiv (-g(x))^{\frac{1}{4}} G_{m,\xi}(x,x'), \tag{4.44}$$

which satisfies

$$(\Box_\eta - (m^2 + (\xi - \frac{1}{6})R(x'))\mathbb{G}_{m,\xi}(x,x') \approx -\delta(x-x'), \tag{4.45}$$

where $\Box_\eta \equiv \eta^{ab}\nabla_a\nabla_b$ and acts on the x argument. When the Compton wavelength of the particle is much smaller than the curvature scale, $m^2 \gg \xi R$, an expansion in momentum space yields

$$\mathbb{G}_{m,\xi}(p) \approx \frac{1}{p^2+m^2} - (\xi - \frac{1}{6})R(x')\frac{1}{(p^2+m^2)^2}, \tag{4.46}$$

where the first term is the flat momentum space Green's function, which is independent of ξ. Its spacetime Fourier transform

$$G_{m,\xi}(x,x') \approx \left(1+\frac{1}{12}R_{ab}(x')x^a x^b\right)G_m^{(\eta)}(x,x') + \frac{1}{2m}(\xi-\frac{1}{6})R(x')\,\partial_m G_m^{(\eta)}(x,x'), \tag{4.47}$$

where $G_m^{(\eta)}(x,x')$ is the massive Green function in $\mathbb{M}^d$.

In $d=2,4$,

$$\begin{aligned} G^{(2)}_{m,\xi}(x,x') \approx& \theta(x_0)\theta(\tau^2)\left[\frac{1}{2}J_0(m\tau)+\frac{R(x')\tau^2}{48}J_2(m\tau)-\frac{\xi R(x')\tau}{4m}J_1(m\tau)\right] \\ G^{(4)}_{m,\xi}(x,x') \approx& \theta(x_0)\left[\left(\frac{1}{2\pi}\delta(\tau^2)-\theta(\tau^2)\frac{m}{4\pi\tau}J_1(m\tau)\right)\left(1+\frac{1}{12}R_{ab}(x')x^a x^b\right)\right] \\ &-\theta(x_0)\theta(\tau^2)\left(\xi-\frac{1}{6}\right)\frac{R(x')}{8\pi}J_0(m\tau). \end{aligned} \tag{4.48}$$

This spacetime form makes it possible to look for causal set counterparts.

We begin with defining the massless Green's functions and then generating the massive one using Eq. (4.5) [157,158]. To this first order approximation, $R(x')$ is a constant, so not only can $G_{m,\xi}(x,x')$ be obtained from $G_{0,\xi}(x,x')$, but also any other $G_{m',\xi'}(x,x')$.

Since we already know the causal set Green's functions in $\mathbb{M}^{2,4}$ one can first attempt to use them to generate other Green's functions. For any pair of conformally related spacetimes $g_{ab}=\Omega^2 g'_{ab}$, the conformally coupled ($\xi_c=\frac{d-2}{4(d-1)}$) massless Green's functions satisfy

$$G_{0,\xi_c}(x,x') = \Omega^{\frac{2-d}{2}}(x)G'_{0,\xi_c}(x,x')\Omega^{\frac{2-d}{2}}(x'). \tag{4.49}$$

This suggests a way of obtaining the causal set Green's function on spacetimes that are conformally flat.

Any $d=2$ spacetime is locally conformally flat and the conformal coupling $\xi_c=0$. Therefore, as in $\mathbb{M}^2$, the massless causal set Green's function is

$$K^{(2)}_{0,0}(e,e') = \frac{1}{2}C(e,e') \tag{4.50}$$

for any curved spacetime, not only in the RNN. Using Eq. (4.5) for any spacetime in $d=2$

$$\begin{aligned} K^{(2)}_{m,0}(e,e') &= \sum_{k=0}^{\infty}\left(-\frac{m^2}{\rho}\right)^k\left(\frac{1}{2}\right)^{k+1}\underbrace{\left(C*C*\ldots C\right)}_{k+1}(e,e') \\ &= \sum_{k=0}^{\infty}\left(-\frac{m^2}{\rho}\right)^k\left(\frac{1}{2}\right)^{k+1}C_{k+2}(e,e'). \end{aligned} \tag{4.51}$$

Thus,

$$\langle \mathbf{K}^{(2)}_{m,0}(x,x')\rangle = G_{m,0}(x,x'). \tag{4.52}$$

In the RNN, the curvature is constant to leading order and hence we can replace m by the generalised mass term $M = \sqrt{m^2 - \xi R}$ in the above expression. Thus, the order theoretic form of the causal set Green's function does not change in $d = 2$ with the addition of curvature.

In the RNN it is possible to explicitly verify this relation [157,158] since, as shown in Sect. 3.3, [115]

$$\langle \mathbf{C}_k(x,x')\rangle \approx \langle \mathbf{C}_k(x,x')\rangle_\eta \left(1 - \frac{R(x')\tau^2}{24}\frac{k}{k+1}\right). \tag{4.53}$$

Exercise 4.5 Using Eqs. (4.53) and (4.48) show that Eq. (4.52) is satisfied.

The $d = 4$ RNN Green's function in Eq. (4.48) is much more complicated and it is challenging to guess its order theoretic form. We instead ask under what conditions $K^{(4)}_{0,\xi}$ is proportional to the link matrix L, as for example in Eq. (4.18). This simplifies the calculation to the evaluation of $\langle \mathbf{L}(x,x')\rangle = \exp(-\rho V(x,x'))$ in the RNN. Using the expansion [18,120,121]

$$V(x,x') \approx V_\eta(x,x')\left(1 - \frac{1}{180}R(x')\tau^2 + \frac{1}{30}R_{ab}(x')x^a x^b\right), \tag{4.54}$$

where $V_\eta(x,x') = \frac{\pi}{24}\tau^4(x,x')$,

$$\lim_{\rho\to\infty}\frac{\sqrt{\rho}}{2\pi\sqrt{6}}\langle \mathbf{L}(x,x')\rangle \approx \frac{1}{2\pi}\theta(x_0)\delta(\tau^2)\left(1 - \frac{R_{ab}(x')x^a x^b}{60}\right). \tag{4.55}$$

since the $\delta(\tau^2)\tau^2$ term vanishes, and we have used

$$\begin{aligned}\lim_{\rho\to\infty}\sqrt{\rho}\exp(-c\rho\tau^4) &= \frac{\Gamma(1/2)}{2\sqrt{c}}\delta(\tau^2),\\ \lim_{\rho\to\infty}\sqrt{\rho}(c\rho\tau^4)\exp(-c\rho\tau^4) &= \frac{\Gamma(3/2)}{2\sqrt{c}}\delta(\tau^2).\end{aligned} \tag{4.56}$$

Comparing with Eq. (4.48) it is clear that the ansatz for the causal set Green's function in $\mathbb{M}^4$ does not in general produce the continuum Green's function in the RNN.

However, there are special cases in which it does [157,158]. In Ricci-flat spacetimes $R_{ab} = 0$, the lowest order corrections to the volume Eq. (4.54) vanish which implies that to leading order the identification $K^{(4)}_{0,\xi}(e,e') = \frac{\sqrt{\rho}}{2\pi\sqrt{6}}L(e,e')$ gives the continuum Green's function $G^{(4)}_{0,\xi}(x,x')$ to this order, for any ξ (so long as the Compton condition is satisfied) in the limit $\rho \to \infty$. We can thus generate $K^{(4)}_{m,\xi}(x,x')$ using

Eq. (4.5), and to this order $\langle \mathbf{K}^{(4)}_{m,\xi}(x, x')\rangle$ limits to $G^{(4)}_{m,\xi}(x, x')$. This form could therefore be useful in studying the UV behaviour in an RNN of a Ricci flat spacetime like Schwarzschild.

For Einstein spaces, $R_{ab} \propto g_{ab}$, and with conformal coupling $\xi_c = \frac{1}{6}$, the terms $\delta(\tau^2)g_{ab}x^a x^b$ in Eqs. (4.48) and (4.55) vanish, which again allows us to identify $K^{(4)}_{0,\frac{1}{6}}(e, e') = \frac{\sqrt{\rho}}{2\pi\sqrt{6}} L(e, e')$. This in turn can be used to generate $K^{(4)}_{m,\frac{1}{6}}(x, x')$ using Eq. (4.5).

4.2.3 Discrete Green's Functions on De Sitter and Anti De Sitter Spacetimes

Both de Sitter (dS_4) as well as anti-de Sitter (adS_4) spacetimes are of constant curvature, and hence we can use the effective mass M when generating the massive from the massless Green's functions [157, 158] (Fig. 4.3).

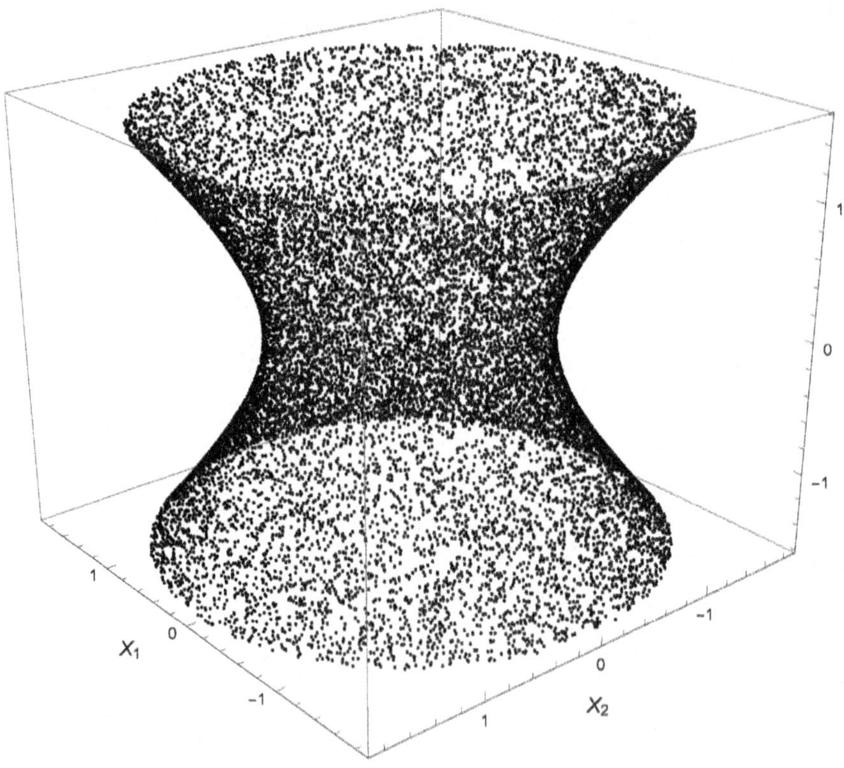

Fig. 4.3 A Poisson sprinkling into a portion of $d = 2$ de Sitter spacetime embedded in $\mathbb{M}^3$. The relations on the elements are deduced from the causal structure of $\mathbb{M}^3$

The conformally flat dS4 and adS4 line elements can be written as

$$ds_{\mathrm{dS}}^2 = \frac{1}{(1+Hx_0)^2}\left(-dx_0^2 + \sum_{i=1}^{3} dx_i^2\right),$$

$$ds_{adS}^2 = = \frac{1}{(1+Hx_3)^2}\left(-dx_0^2 + \sum_{i=1}^{3} dx_i^2\right), \tag{4.57}$$

where $H = \sqrt{\frac{\Lambda}{3}}$. In dS4 the conformal time x_0 satisfies $-\frac{1}{H} < x_0 < \infty$, while in adS4 we have made the shift $x_3 \to x_3 + \frac{1}{H}$ from the standard form. Using

$$G_{0,\frac{1}{6}}(x,x') = \Omega^{-1}(x) G_0^{\eta}(x,x') \Omega^{-1}(x'), \tag{4.58}$$

we see that

$$G_{0,\frac{1}{6}}^{(\mathrm{dS})}(x,0) = \frac{1}{2\pi}\theta(x_0)\delta(\tau^2(x,x'))(1+Hx_0),$$

$$G_{0,\frac{1}{6}}^{(\mathrm{adS})}(x,0) = \frac{1}{2\pi}\theta(x_0)\delta(\tau^2(x,x'))(1+Hx_3). \tag{4.59}$$

where we have taken x' to lie at the origin. This is possible because of spacetime homogeneity.

Since $G_0^{\eta}(x,x')$ is related to the flat spacetime link matrix this is again a good place to start. Comparing with Eq. (4.16)

$$\lim_{\rho\to\infty}\sqrt{\frac{\rho}{\pi}}\langle \mathbf{L}(x,x')\rangle = \theta(x_0)\theta(\tau^2(x,x'))\delta(\sqrt{V(x,x')}). \tag{4.60}$$

The calculation then reduces to evaluating $V(x,x')$ in dS4 and adS4. We can however simplify the calculation by noticing that the contribution to $\langle \mathbf{L}(x,x')\rangle$ for a fixed x is dominated by those x' which lie near the past light cone of x, i.e., for small $V(x,x')$.

In (x_0, x_1, x_2, x_3) coordinates we define the light cone coordinates

$$u = \frac{1}{2}(x_0 - x_3),\ v = \frac{1}{2}(x_0 + x_3). \tag{4.61}$$

While we know its form in the RNN, we also need to take into account contributions from "long skinny" neighbourhoods, which lie well outside of the RNN. Such a neighbourhood is an Alexandrov interval for which one of these lightcone coordinates ranges over a small ϵ and the other over some $L \gg \epsilon$. Taking x' to be at the origin, let $u(x) = \epsilon$, $v(x) = L$, with $\alpha^2 \equiv \frac{\epsilon}{L} \ll 1$. In this approximation, the distinction between dS4 and adS4 disappears since

$$1 + Hx_0 = 1 + H(L+\epsilon) \approx 1 + HL, \quad 1 + Hx_3 = 1 + H(L-\epsilon) \approx 1 + HL. \tag{4.62}$$

The spatial rotational symmetry means we need consider only the $x_0 - x_3$ plane, and moreover, perform a boost with parameter $\beta = \frac{x_3}{x_0} \approx 1 - 2\alpha$ about x' in the $x_0 - x_3$ plane which brings us to $x = (\tilde{x}_0, 0, 0, 0)$. In these coordinates, for both dS_4 and adS_4, the conformal factor at an event x'' in the interval is

$$\widetilde{\Omega}^2(x'') \approx \frac{1}{(1 + A(x_0'' + x_3''))^2} \tag{4.63}$$

where $A = \frac{1}{2}H\alpha$. Using $\tau^2 = 4L\epsilon$, we find

$$\sqrt{V(x,x')} = \frac{1}{2}\sqrt{\frac{\pi}{6}}\frac{\tau^2}{(1 + A\tau)} \approx \frac{1}{2}\sqrt{\frac{\pi}{6}}\left(\frac{4\,L\epsilon}{1 + HL}\right), \tag{4.64}$$

and hence

$$\lim_{\rho\to\infty}\sqrt{\frac{\rho}{\pi}}\langle \mathbf{L}(x,x')\rangle = \theta(x_0 - x_0')\theta(\tau^2(x,x'))\sqrt{\frac{6}{\pi}}(1 + HL)\delta(4\,L\epsilon). \tag{4.65}$$

Since the continuum Green's function has support only on the light cone on which $1 + Hx_0 = 1 + Hx_3 = 1 + HL$,

$$G_{0,\frac{1}{6}}(x,x') \approx \frac{1}{2\pi}(1 + HL)\theta(\tau^2)\theta(x_0)\delta(4L\epsilon). \tag{4.66}$$

which implies that

$$\lim_{\rho\to\infty}\frac{1}{2\pi}\sqrt{\frac{\rho}{6}}\langle \mathbf{L}(x,x')\rangle = G_{0,\xi_c}(x,x'). \tag{4.67}$$

Thus, we find that the causal set Green's functions on the conformal patches of dS_4 and adS_4 are also of the form

$$K^{(dS_4,adS_4)}_{0,\frac{1}{6}}(e,e') = \frac{1}{2\pi}\sqrt{\frac{\rho}{6}}L(x,x') \tag{4.68}$$

with $K^{(dS_4,adS_4)}_{m,\xi}$ given by Eq. (4.5). Because the conformal patch in dS_4 is causally complete, this result extends to global dS_4 as well [157,158].

A few words are in order regarding causal sets sprinkled into global adS_d. Because of its timelike conformal boundary, all causally related pairs x, x' which do not lie in the same conformal patch of adS_d have Alexandrov intervals $I(x,x')$ of infinite volume, even though $\tau(x,x')$ is finite. Hence a Poisson sprinkling into a chunk of adS_d larger than the conformal patch gives rise to posets which are not locally finite (and hence not technically causal sets.) Our restriction to the conformal patch on the other hand means that the events in the Poisson sprinkling satisfy the condition of local finiteness, and hence are causal sets.

Another interesting feature we notice when sprinkling into a finite region in the conformal patch of adS_d is a strange manifestation of "holography" due to the fact that the conformal factor grows with the spatial radius. This means that most of the elements of a causal set sprinkled into a finite radius R region of adS_d (within the conformally flat patch) are concentrated on the boundary at R, thus mimicking adS_d holography! Whether these oddities warrant further investigation is a matter of taste—the physical relevance of adS_d is a debate which we will not enter into here.

4.3 The Sorkin-Johnston Construction

Having obtained the retarded and advanced Green's functions, we now turn to quantising the scalar field. The standard textbook approach to quantising a scalar field in (M, g) starts with the equal time commutation relations for the quantum field operators. As discussed in various contexts in this book, the discrete analogues of an inextendible spatial hypersurface is an inextendible antichain which, however, is not Cauchy. Instead we need a covariant, spacetime algebra, provided by the spacetime commutation relation or **Peierls bracket**

$$[\Phi(x), \Phi(x')] = i(G_R(x, x') - G_A(x, x')) \equiv i\Delta(x, x'), \tag{4.69}$$

where $G_{R,A}(x, x')$ are the retarded and advanced Green's functions and where $i\Delta(x, x')$ is the **Pauli Jordan function**. The insight gained by Johnston [38,39] is that the covariantly defined image of $i\Delta(x, x')$ can be used to define the quantum field vacuum. This can be traced to the following result in the continuum, i.e., $Im(i\Delta) = ker(\Box)$ where $\Box$ denotes the continuum Klein Gordon, or scalar field D'Alembertian. Thus, the space of solutions for the Klein Gordon field is the same as the image space of $i\Delta$. Sorkin showed that this procedure extends to the continuum and can be used to define a preferred and yet covariant (observer-independent) vacuum in any continuum spacetime [156,168]. Because of its operator theoretic underpinnings, this **Sorkin-Johnston** (SJ) vacuum has been studied (and critiqued) by the AQFT community [159,169,170]. In particular, it was shown that the unique vacuum is not Hadamard. However, given the underlying discreteness and the changed UV structure that CST brings with it, this may simply be a statement of new, yet unexplored UV physics.

The Pauli Jordan (PJ) function $i\Delta(x, x')$ is both antisymmetric, $i\Delta(x, x') = -i\Delta(x', x)$ and purely imaginary. It is therefore a Hermitian matrix over the elements of the causal set. In the continuum, it is the kernel of the self-adjoint integral operator

$$i\widehat{\Delta} \circ f(X) \equiv \int_R i\Delta(X, X') f(X')\, dV_{X'}, \tag{4.70}$$

where f belongs to the space of $\mathcal{L}^2$ integrable functions on a bounded, globally hyperbolic region $R \subset (M, g)$, with inner product

$$(f, g) \equiv \int_M dV_x f^*(x) g(x). \tag{4.71}$$

The eigenvalues of $i\widehat{\Delta}$ are therefore real and because $\Delta(x, x') = -\Delta(x', x)$ they also come in positive and negative pairs

$$\begin{aligned} i\widehat{\Delta} \circ u_k &= \lambda_k u_k \\ i\widehat{\Delta} \circ u_k^* &= -\lambda_k u_k^*, \end{aligned} \tag{4.72}$$

where $u_k \in \mathrm{Im}(i\widehat{\Delta})$. The normalised **SJ modes** $u_k^{SJ} = \sqrt{\lambda_k} u_k$ form a complete orthonormal basis in $\mathrm{Im}(i\widehat{\Delta})$ with respect to the $\mathcal{L}^2$ inner product on R. This gives a spectral decomposition of the PJ operator

$$i\Delta(x, x') = \sum_k \lambda_k \left(u_k(x)u_k^*(x') - u_k^*(x)u_k(x')\right). \tag{4.73}$$

In the continuum, it is known that

$$\mathrm{Im}(i\widehat{\Delta}) = \ker(\Box), \tag{4.74}$$

where $\Box$ is the Klein-Gordon operator. Hence the the SJ modes are also solutions of the KG equation. Since the SJ eigenvalues come in positive and negative pairs, (u_k, u_k^*) can be used to define the quantum field operator

$$\widehat{\Phi}(x) \equiv \sum_k \sqrt{\lambda_k} \left(\mathbf{a}_k u_k(x) + \mathbf{a}_k^\dagger u_k^*(x)\right), \tag{4.75}$$

where $\mathbf{a}^\dagger{}_k$, $\mathbf{a}_k$ are the SJ creation and annihilation operators, with $\mathbf{a}_k|0\rangle_{SJ} = 0$ defining the SJ vacuum. This is an observer independent notion of positive and negative frequencies and hence of the quantum field theory vacuum $|0\rangle_{SJ}$. The only caveat is that the region R is of finite volume, though even this can be relaxed for static spacetimes with compact Cauchy hypersurfaces (see [171].) It follows that the Wightmann operator associated with the two point Wightmann function

$$\begin{aligned} W_{SJ}(x, x') &\equiv \langle \Phi(x), \Phi(x')\rangle, \\ \widehat{W}_{SJ} = \mathrm{Pos}(i\widehat{\Delta}) \Longleftrightarrow \widehat{W}_{SJ} &= \frac{i}{2}\widehat{\Delta} + \sqrt{-\widehat{\Delta}^2}, \\ \Rightarrow W_{\mathrm{SJ}}(x, x') &= \sum_k \lambda_k u_k(x) u_k^*(x'). \end{aligned} \tag{4.76}$$

It is this form that we will find most useful when quantising the field on the causal set.

On a finite causal set C the operator $i\widehat{\Delta}$ is defined as

$$i\widehat{\Delta} \circ f(e) = i \sum_{e' \in C} \Delta(e, e') f(e') \tag{4.77}$$

where $f : C \to \mathbb{C}$. The operator $i\widehat{\Delta}$ is therefore Hermitian on a finite C, and hence self-adjoint, with eigenfunctions (u_k, u_k^*) associated with the pair of eigenvalues $(\lambda_k, -\lambda_k)$. The spectral decomposition of the PJ operator is

$$i\Delta(e, e') = \sum_k \lambda_k \left(u_k(e)u_k^*(e') - u_k^*(e)u_k(e')\right), \tag{4.78}$$

and hence

$$W_{SJ}(e, e') = \sum_k \lambda_k u_k(e)u_k^*(e') = \mathrm{Pos}(i\Delta(e, e')). \tag{4.79}$$

A generic state $\widehat{W}$ is given by

$$W(e, e') = R(e, e') + \frac{i}{2}\Delta(e, e'), \tag{4.80}$$

where $R(e, e')$ is a real and symmetric matrix, while for the SJ state, $R(e, e') = \sqrt{-\Delta^2}(e, e')$.

Exercise 4.6 Calculate the $d = 2$ and $d = 4$ massless scalar two point functions on the causal sets in Fig. 4.4.

For continuumlike causal sets $W_{SJ}(e, e')$ can only be calculated numerically by first finding the SJ modes of the discrete $i\widehat{\Delta}$. Simulations show that the spectrum of $i\widehat{\Delta}$ exhibits an interesting feature in the extreme UV [41,43,164]. We expect, of course, that the spectrum has a maximum eigenvalue k since causal sets provide a natural covariant UV cut-off, but the detailed UV behaviour exhibits significant deviation from the continuum after a characteristic scale [171]. As we will see in Sect. 4.4, this deviation becomes important in the calculations of the SSEE.

Consider the massless scalar field on $\mathbb{M}^2$. In the continuum, the SJ modes for the causal diamond $\mathbb{D}^2$ come in two sets [39,168]

$$\begin{aligned} f_k(u, v) &= e^{iku} - e^{ikv}, \quad k = n\pi/L, \\ g_\kappa(u, v) &= e^{i\kappa u} + e^{i\kappa v} - 2\cos(\kappa L), \quad \tan(\kappa L) = 2\kappa L, \end{aligned} \tag{4.81}$$

where $2\sqrt{2}L$ is the proper time in $\mathbb{D}^2$. The first set $\{f_k(u, v)\}$ has eigenvalues $\lambda_k = \pm L/k$ and is antisymmetric with respect to $u \leftrightarrow v$, while the second set $\{g_\kappa(u, v)\}$ has eigenvalues $\pm\lambda_\kappa L/\kappa$, and is symmetric. For large values $\kappa \sim (2n + 1)\pi/L$ [172],

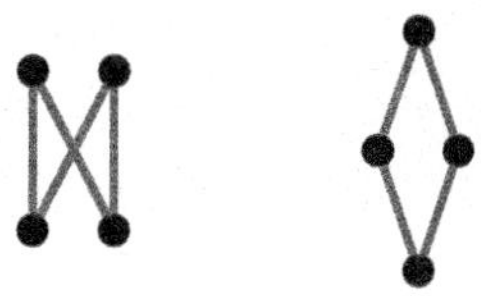

Fig. 4.4 Causal sets for Exercise 4.6

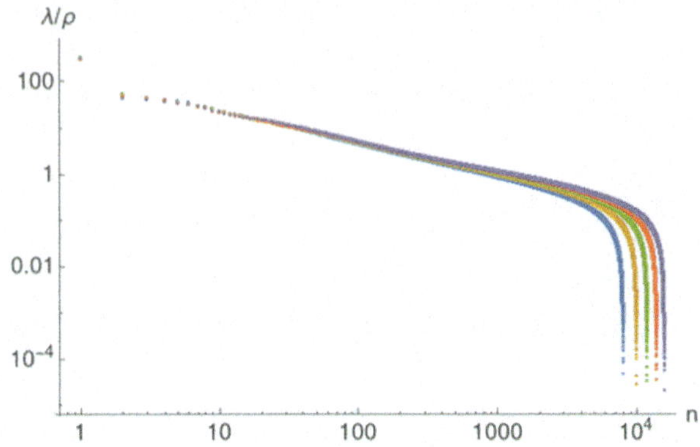

Fig. 4.5 A log-log plot of the positive eigenvalues of $i\Delta$ in dS_4 for different values of ρ. There is a scaling behaviour upto a knee, beyond which the spectrum exhibits a non-scaling behaviour. As ρ increases, the knee shifts to larger n values [174]

so we can collectively refer to both sets of modes using a common n-label. Thus generically for large n, λ_n scales inversely with n (see [173] for the massive case.)

When we plot the SJ eigenvalues for causal sets in $\mathfrak{C}_\rho(\mathbb{M}^2)$, we find a strikingly different behaviour for large n as shown in Fig. 4.5. As expected, in the causal set the spectrum terminates at some finite n_{max}, and while it tracks the continuum scaling behaviour for low n values, there is a sharp transition, or "knee" in the log-log spectrum, beyond which there is a departure from the scaling behaviour. As the density ρ increases, the knee shifts towards the UV. In [171] it was shown that the functional form for the discrete spectrum satisfies a scaling behaviour upto the knee, but veers sharply away from this behaviour beyond the knee. It can be modelled by the function

$$\lambda \equiv \begin{cases} \frac{\beta_1}{n^{\alpha_1}}, & \text{for } n < n_0 \\ \alpha_2 n + \beta_2, & \text{for } n > n_0, \end{cases} \tag{4.82}$$

where $\beta_2 > 0$, $\alpha_2 < 0$ which sets a minimum scale $\lambda_m = \frac{\beta_2}{|\alpha_2|}$. This modification of the spectrum is very interesting because it tells us that rather than becoming more and more like $\mathbb{M}^2$, as one expects in continuum quantum field theory, there is a scale beyond which new physics manifests itself. How this affects the RG equations for quantum field theory is therefore of interest, and suggests new, beyond standard model phenomenology.

Exercise 4.7 Calculate the spectrum of $i\widehat{\Delta}$ in the causal sets of Fig. 4.4, using the $d = 2$ and the $d = 4$ Green's functions. Does the kernel of $i\Delta$ change depending on this choice? In general, which kernel do you think will be bigger?

In Figs. 4.6 and 4.7 we show results from simulations of $W_{\mathrm{SJ}}(e, e')$ [41] for the massless free scalar field for causal sets $C \in \mathfrak{C}_\rho(\mathrm{dS}_{2,4})$ [41, 158].

For $\mathfrak{C}_\rho(\mathrm{dS}_2)$ the causal set W_{SJ} shows very good agreement with the Bunch Davies vacuum as seen in the scatter plots Fig. 4.6. However, the causal set W_{SJ} for dS_4 [41] shows a significant deviation from the Bunch Davies vacuum for the small

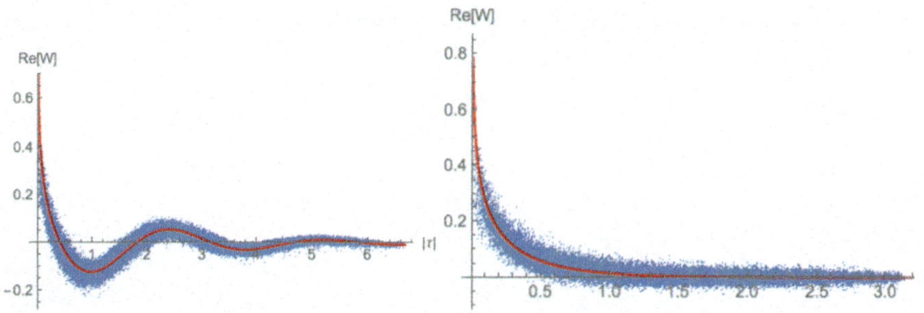

Fig. 4.6 Scatter plots of the real part of the causal set W_{SJ} for $\mathfrak{C}_\rho(\mathrm{dS}_2)$ for $m > 0$ as a function of the proper time τ and the proper distance $|d|$. The line in red shows the dS_2 Bunch Davies vacuum

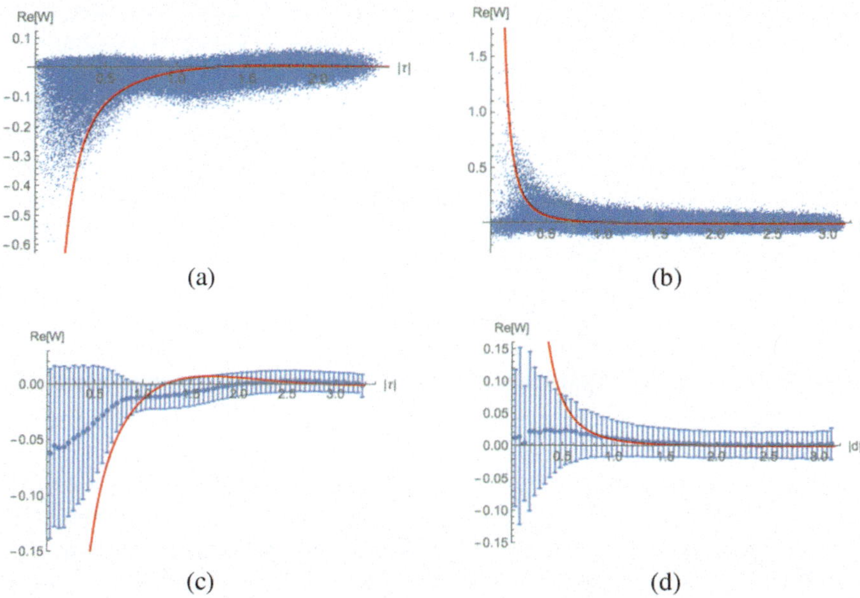

Fig. 4.7 Scatter plots of the real part of the causal set W_{SJ} for $\mathfrak{C}_\rho(\mathrm{dS}_4)$ for $n = 16,000$ elements and $m > 0$, as a function **a** of the proper time τ and **b** the proper distance $|d|$. The line in red shows the dS_4 Bunch Davies vacuum. The causal set shows significant deviations from the continuum at short distances and times. The binned averages and the associated errors with respect to the Bunch Davies vacuum are shown in **c** and **d**

proper time and distance regime, as shown in the scatter and error plots in Fig. 4.7. This can be interpreted as a modification in the quantum field theory vacuum in the extreme UV. This could have interesting cosmological significance, for example in inflationary scenarios.

4.4 Sorkin's Spacetime Entanglement Entropy

Black hole and cosmological horizons have proved to be an extremely rich arena in which quantum effects of geometry manifest themselves, however subtly. The analogy between classical black hole mechanics and thermodynamics is very intriguing, and there are many hints that this is more than an analogy, because of the interplay between geometry and quantum field theory [12,175,176]. However strong these hints, it is still important to look for these effects in a theory of quantum gravity, to establish whether they are realised in full, or are only emergent in the low energy regime in which they have been studied most carefully.

One of the big pieces of the geometry-thermodynamics jigsaw puzzle is black hole or, more generally, horizon entropy. Classically, the geometric area law says that the area of a black hole can never decrease as long as the cosmic censorship conjecture holds and appropriate energy conditions are satisfied [82]. This is analogous to the second law of thermodynamics, which states that the entropy of an isolated system cannot decrease. If area is entropy, then one must ask what the quantum origin of this entropy is or, in common-speak, what the microstates of the black hole are. There have been many answers to this question from string theory, loop quantum gravity and, more recently, causal set kinematics, amongst others [144,145]. If area is indeed entropy, then the process of area increase is equivalent to entropy production, which means it is not reversible. This leads to a second important quandary, which is whether processes in quantum gravity are necessarily unitary. This is exemplified also in the "information loss" puzzle which has been discussed extensively in the literature.

Instead of looking for a microstate origin for horizon entropy, one can also ask if the entropy of the horizon can be a result of the entanglement between the quantum state in the outside region of the black hole with that of the black hole interior [177]. Associated with any such partitioning of a quantum system is the von-Neumann entanglement entropy (EE). In [177] it was shown that the EE obeys an area law suggesting that it could be a source of black hole entropy, an idea that has led to a rich literature spanning black hole physics and quantum information. However, this EE calculation gives rise to a UV divergence in the continuum—to make sense, it needs a UV cut-off to establish the area-entropy relationship. It is therefore of particular interest to study the EE for horizons in causal set theory which comes with a natural, covariant UV cut-off.

We are however already stuck at the beginning—what is an intrinsic definition of a horizon in a causal set? Horizons are sets of measure zero in a Poisson sprinkling and do not make an appearance in continuumlike causal sets. Given that there is only one causal relation $\prec$ it is nontrivial to deduce a null-relation using a second chronological relation, as in Penrose and Kronheimer's axiomatic framework [20]. In [145] it was shown that the number of horizon molecules straddling the intersection of a null hypersurface with a spatial hypersurface is proportional to the area of the intersection. However the definition is not intrinsic to the causal set since one needs the continuum to define the spacelike and null hypersurfaces (see [166] for some recent ideas on this.)

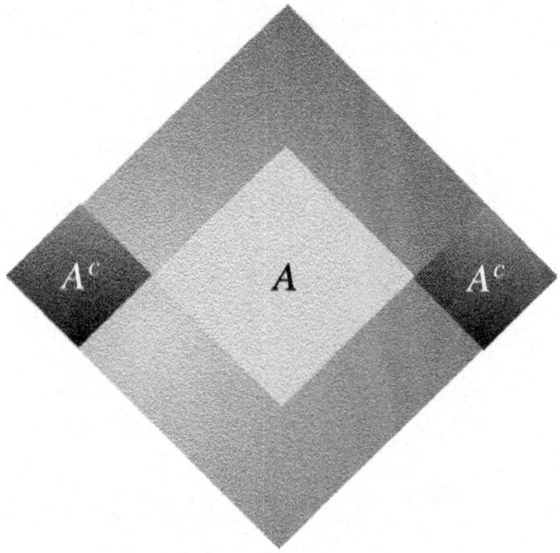

Fig. 4.8 A is a causal diamond nested inside a larger causal diamond in $\mathbb{M}^2$. The region A^{cc} which is causally complementary to A in the larger diamond consists of two disconnected diamonds as shown

On the other hand, the notion of EE is independent of horizons and needs only an identification of causally complementary regions, which are perfectly well defined in a causal set. For any sub-causal set $A \subset C$, define its causally complementary sub-causal set as

$$A^{cc} \equiv \{e \in C \,|\, \forall\, e' \in A,\ e \parallel e'\}, \tag{4.83}$$

with the order relations on the elements induced by those in C. An example of causally complementary regions in a causal diamond in $\mathbb{M}^2$ is shown in Fig. 4.8.

Exercise 4.8 Let C be the causal set in Fig. 4.9 and A the sub-causal set in grey. Find A^{cc}. Show in general that every $e \in C$ is either in A, A^{cc} or to their common future or past.

The standard formulation of the von-Neumann EE is not of use to causal sets since it requires us to define a state at a moment of time, something, we repeat, is not intrinsic to a causal set. Instead, Sorkin's spacetime formulation of EE (SSEE), based on the field correlators that we have just studied, is what we need [42,43,171]. Let A, $A^c \subset C$ be two causally complementary sub-causal sets in C, and let $\widehat{W}_{SJ}$ denote the scalar field SJ on C. Consider its restriction $\widehat{W}_{SJ}|_A$ to $A \subset C$, which is

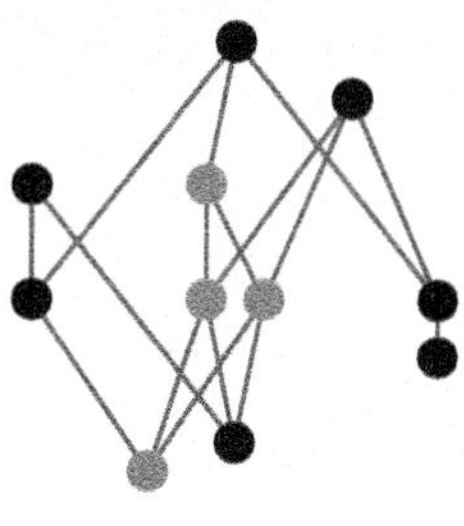

Fig. 4.9 The sub-causal $A \subset C$ is marked in gray

in general a mixed state.[1] The SSEE of the region A is defined as

$$\mathcal{S} = \sum_{\mu} \mu \ln |\mu|, \tag{4.84}$$

where μ are the eigenvalues of the generalised eigenvalue equation

$$\sum_{e' \in A} W(e, e')|_A \Psi(e') = i\mu \sum_{e' \in A} \Delta(e, e')\Psi(e'), \quad \Psi \notin \ker i\Delta, \quad e, e' \in A. \tag{4.85}$$

Because of the finiteness of the matrices involved, it should be obvious that the SSEE is finite.

Exercise 4.9 Let C be the causal set in Fig. 4.9 and let A be the sub-causal set marked in gray. Find the SSEE for A with respect to A^{cc} using the form of the retarded Green's functions for $\mathbb{M}^2$ as well as that for $\mathbb{M}^4$.

The causal set SSEE was first calculated in [43] for complementary order intervals in $C \sim_\rho \mathbb{D}^2$, Fig. 4.10. The surprise in the calculation is that rather than the expected $d = 2$ "area" dependence, i.e., the leading order logarithmic law, the causal set SSEE obeys a volume law [43]! As shown by Sorkin and Yazdi [43] the expected logarithmic behaviour emerges only when a double truncation at the knee of the SJ spectrum is performed while calculating W_{SJ} in the larger diamond as well as $\widehat{\Delta}$ in A. This is also the case in the dS_2, dS_4 calculations of the SSEE performed in [44, 158], shown in Figs. 4.11 and 4.12.

It is a puzzle to figure out what role the additional UV truncation plays in an already UV finite theory. What is clear from the SJ spectrum of Fig. 4.5 is that the

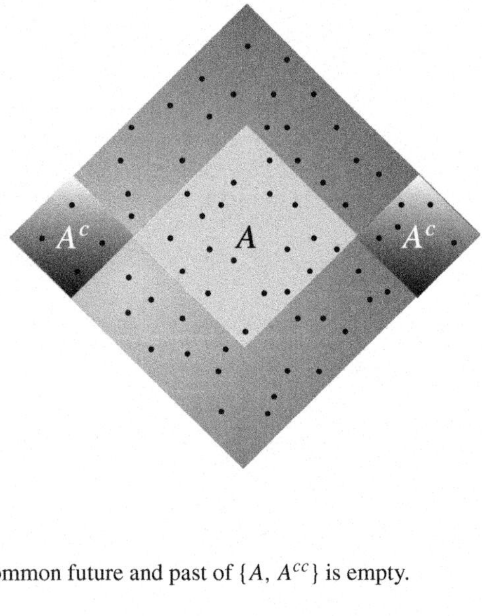

Fig. 4.10 A pair of nested causal sets $C' \subset C$, where C is approximated by a causal diamond in $\mathbb{M}^2$ and C' by a smaller causal diamond. The SSEE is calculated for the scalar field state restricted to C', with respect to the state in C

[1] The restriction of $\widehat{W}|_A$ remains pure if the common future and past of $\{A, A^{cc}\}$ is empty.

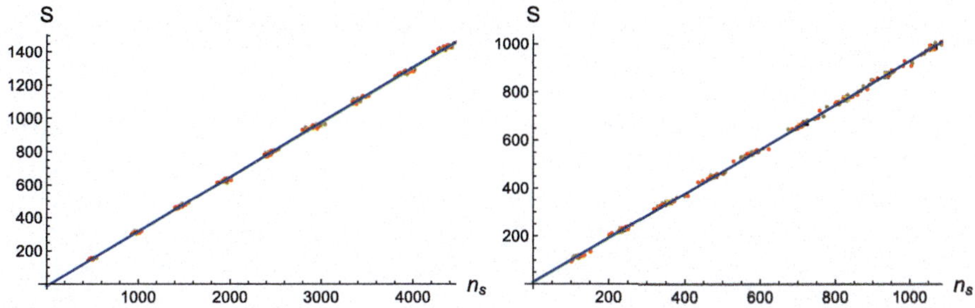

Fig. 4.11 In both dS_2 and dS_4, the SSEE obeys a volume law when there is no truncation of the spectrum

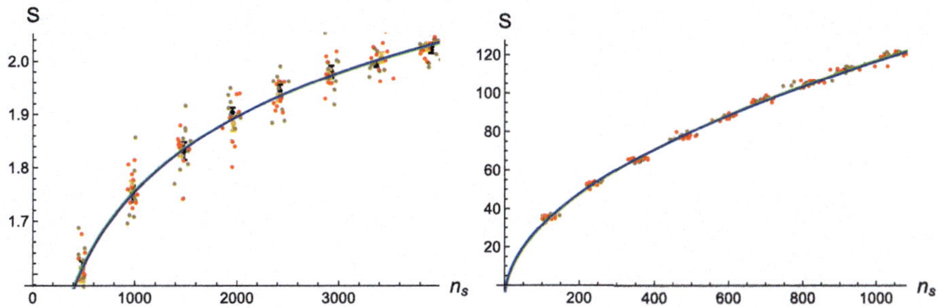

Fig. 4.12 The double truncation procedure in both dS_2 and dS_4, ensures that the SSEE satisfies an area law

truncation procedure of Sorkin and Yazdi occurs at the knee of the SJ spectrum. This is a meso-scale at which local quantum field theory is still valid. At this scale, the SSEE gives rise to an area law as expected. However, beyond this, in the deep UV, the intrinsic causal set nonlocality kicks in. The standard arguments for an emergent area law thus breaks down, as it does in condensed matter systems with long-range or nonlocal interactions, which also exhibit volume rather than area laws [178,179]. This deep UV behaviour of the causal set SSEE may be an important peek into the nature of black hole physics in quantum gravity. It certainly suggests caution when using ideas from black hole thermodynamics at the most fundamental level (see [180] for a very interesting discussion on this issue.)

4.5 Summary

In this chapter we have laid down the basics for constructing a free quantum scalar field on causal sets with explicit constructions for causal sets approximated by $\mathbb{M}^d$, RNNs, dS_d and adS_d spacetimes, for $d = 2, 4$. There has been very interesting recent progress on perturbative quantum field theory using AQFT (pAQFT) techniques and implementing these on causal sets [45–47]. Just as the deep UV regime of the SJ spectrum is seen to modify the EE into a volume law we also expect that standard

RG methods will need to be modified to accommodate the intrinsic nonlocality of causal sets in the deep UV.

An interesting direction we have not touched upon here is the use of the inverse nonlocal D'Alembertian we defined in Chap. 3 to construct the free massless scalar quantum field theory [181]. The resulting continuum version of the nonlocal propagator has branch cuts which result in a continuum of off-shell massive modes along with the on-shell massless modes. Thus, while the mode expansion of the quantum field operator includes both sets of modes, the S-matrix amplitudes for interaction between these modes is zero. The off-shell modes can thus be viewed as potential candidates for dark matter [181, 182]. The D'Alembertian has also been used to study the Landau pole problem in scalar field theory [183].

Dynamics

5

Abstract

Because of the inherent nonlocality of continuumlike causal sets, CST dynamics is best suited to the covariant path integral or sum-over-histories framework. The continuum-inspired CST partition function is a discrete sum over causal sets with complex weights given by the discrete BDG action. An important analytic result for the theory is that the BDG action suppresses the most entropically dominant, non-continuumlike contributions for large n, while providing a minimum discreteness scale of order the Planck scale in all dimensions. The partition function can be analytically continued into a theory of Lorentzian statistical geometry and studied numerically, by Markov Chain Monte Carlo techniques. The dimensionally restricted cases in $d = 2$ and $d = 3$ exhibit a first order phase transition between a continuumlike and a noncontinuumlike phase, which can be interpreted in terms of a Hartle-Hawking wave function. A more fundamental dynamics based on purely order theoretic principles is the Rideout-Sorkin sequential growth models in which a causal set is grown element by element. The dynamics is Markovian and satisfies internal temporality, covariance and spectator independence. It is defined by a measure triple (either classical or quantum) where the observables (or beables) are events in a covariant-sigma algebra, some of which have cosmological significance.

5.1 Introduction

The most challenging task in a non-perturbative theory of quantum gravity is to construct a consistent, covariant quantum dynamics of the kinematical degrees of freedom. CST is not amenable to canonical methods because it lacks the equivalent of a Cauchy hypersurface and is hence more suited to the broad, histories-based framework for quantum theory. In this chapter we will use both the discrete path integral or path sum, as well as the double path integral or decoherence functional, to define the dynamics for causal sets.

S. Surya, *The Causal Set Approach to Quantum Gravity*, Lecture Notes in Physics 1036,
https://doi.org/10.1007/978-3-031-84420-1_5

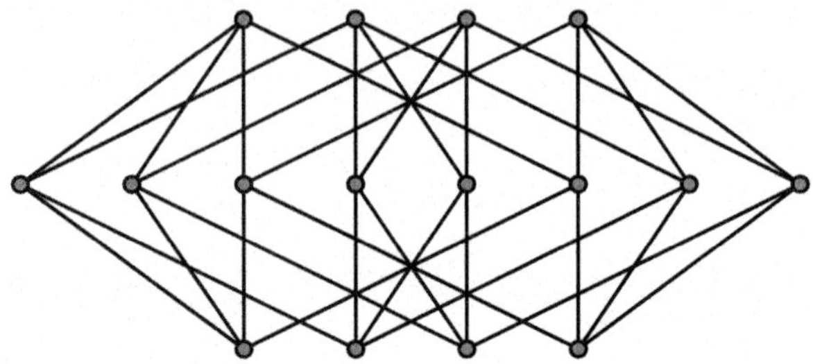

Fig. 5.1 An example of a KR order

In Chap. 3 we described the "first-quantisation" process of extracting a ensemble of causal sets underlying a given spacetime. This gives us a class of continuumlike causal sets. We now have the much harder task of "second-quantisation", with the path sum including all possible causal sets, not just those that are continuumlike. Unlike the continuum path integral where one integrates over the space of all four dimensional Lorentzian geometries in the continuum path integral (and possibly sums over all topologies), one must consider *all* causal sets in the CST path-sum, including, but definitely not limited to, those approximated by spacetimes of *any* dimension. What does this **sample space** Ω of all causal sets look like?

While Ω has been explicitly enumerated only upto $n = 16$ (see Appendix A) the space of n-element causal sets Ω_n is known to grow super-exponentially with n, i.e., $|\Omega_n| \sim 2^{\frac{n^2}{4}+\frac{3n}{2}+o(n)}$. Importantly, as shown by Kleitmann and Rothschild [20] it is overwhelmingly dominated (to the same order) by the 3-layered **KR-posets** which are strictly not continuumlike (see Fig. 5.1.) Subdominant to these are an entire hierarchy of non-continuumlike K-layered posets [21,22,50]. In other words, a typical n-element causal set is nothing like the continuum even at large n. The challenge to CST is therefore significant—the classical limit of a viable quantum dynamics on Ω_n must not only overcome this entropic dominance of non-continuumlike causal sets, but also produce the $d = 4$ dimensional cosmological spacetime we live in. This is a worry shared by other approaches to quantum gravity, an example being the dominating branched polymer configurations in dynamical triangulations [51].

We begin in Sect. 5.2 by defining the partition function for CST which is a sum over Ω_n, the sample space of n-element causal sets, each weighted by a continuum-inspired complex weight determined by the BDG action. In Sect. 5.2.1 we describe an important result on the contribution of the entropically dominant K-layer posets, including the KR posets [52–55]. Using a saddle point approximation for large n, it can be shown that the BDG action, in any dimension, suppresses the contributions from these posets when the discreteness scale ℓ, is larger than some (small) multiple of the Planck scale ℓ_p [52–55].

Gaining a more detailed analytic handle on the path integral is however very challenging, especially in extracting physical information like the expectation values of covariant observables. Inspired by other approaches to quantum gravity like CDTs [3] we analytically continue the path integral from a complex path sum into a real path sum over causal sets. In quantum field theory this is performed via a Wick rotation of the time parameter $t \rightarrow it$, but of course this is not a viable option for generic spacetimes, and definitely not for causal sets, which have no Euclidean counterpart! Instead one can perform an analytic continuation from a complex measure into a real measure, while still summing over the same Lorentzian space Ω_n. The analytic continuation gives us a **Lorentzian statistical model**, or state-sum model which can be studied using powerful numerical methods. In Sect. 5.2.2 we describe Markov Chain Monte Carlo (MCMC) methods to evaluate expectation values for different geometric order invariants. However, the full path sum is still hard to evaluate and we can simplify by considering dimensionally restricted sample spaces. These have been studied extensively in $d = 2$ and $d = 3$ [56,60,61,184,185]. In all these examples one finds a ubiquitous first order phase transition [185] from a continuumlike phase to a highly connected non-continuum or "crystalline" phase, as a function of the analytic continuation parameter β [56,60,61,185].

In approaches to quantum gravity where discreteness is used as a regulator, continuous phase transitions are essential to show a meaningful continuum limit. This is not the case in CST where discreteness is assumed to be fundamental, and hence a first order phase transition is a possible indication of non-trivial physics. A hint that such transitions may play a role in early universe physics comes from a numerical study of the $d = 2$ Hartle-Hawking wave-function ψ_{HH}, obtained from a discrete adaptation of the no boundary proposal [57]. ψ_{HH} shows an interesting behaviour at the first order phase transition – for $\beta < \beta_c$ it is sharply peaked on continuum-like causal sets while for $\beta > \beta_c$ the peak shifts to a highly causally connected and homogenous geometry, which mimics the conditions of our early universe. Whether this is a robust result in higher dimensions remains to be seen.

How fundamental is the the BDG action? At the deep quantum level, far from continuum considerations, it would be natural to start with a more intrinsic, order theoretic action arising from a more basic set of first principles. An example of a purely order theoretic action is the **link action** obtained by simply counting the number of links N_0 in a finite causal set. Any such choice of action is, however, ad-hoc unless supported by (a) a set of basic principles and (b) a demonstration (possibly by renormalisation methods [186]), that the BDG action arises in the low energy limit.

In Sect. 5.3 we describe the **sequential growth** paradigm and a principles based approach to dynamics. In sequential growth the causal set is grown element by element, starting with a single element. This initial condition, based on cardinality, has a particularly charming appeal—what can be a more natural choice than the certainty that one has a single element with which spacetime began? In Sect. 5.3.2 we describe the Rideout-Sorkin classical stochastic sequential growth models where the probability measure is Markovian and satisfies local causality (internal temporality) and discrete covariance (label invariance), as well as spectator independence or a "Bell-causality" condition [62]. From this one can deduce that the transition probability from an n-element "parent" causal set to one of its $n + 1$-element "children" causal sets is determined by a single parameter at every stage n. The growth respects the order relation but produces labelled causal sets. Because the growth occurs to the future, the sample space generated is one made up of **past finite** causal sets.

In Sect. 5.3.1.2 we discuss the broad measure theoretic framework of stochastic dynamics. Covariant observables (or beables) in such a formulation are simply the measurable sets of the dynamics. In Sect. 5.3.3 we describe the quantisation of these stochastic models using the quantum measure or decoherence functional. When it satisfies strong positivity, the quantum measure can be expressed as a vector measure which takes values in a **histories Hilbert space** $\mathcal{H}_{\mathrm{h}}$. In order to construct a fully covariant quantum dynamics, however, the quantum measure must satisfy certain stringent convergence conditions. At present we have results in the simplest case of complex sequential growth [71] where $\mathcal{H}_{\mathrm{h}} \simeq \mathbb{C}$. Generalising to more realistic dynamics is one of the big challenges in CST, and will require a much deeper understanding of quantum dynamics, both technically and conceptually.

We have mentioned observables in passing, but let us pause to understand what CST has to say about the covariant observables of quantum gravity. Guided by the kinematics one knows that classical covariant observables must be constructed from order invariants which are label invariant. Without a Hilbert space based formulation of observables as self adjoint operators, one needs an alternative definition of quantum observables. For the analytically continued partition function of Sect. 5.2.2 the expectation values of kinematical observables are all real, and hence this theory is well defined. For the full Lorentzian path integral of Sect. 5.2, however, we do not have any way at present to show that the expectation values of order invariants is real. A more comprehensive picture of observables can however be formulated in the double path integral approach of Sect. 5.3 where observables are "quantum measurable" covariant events in a sigma algebra [65].

But how should observables or beables be viewed at all in quantum gravity, without recourse to external observers? This gaping question is beyond the scope of this book. A promising approach lies in the general direction of the coevent interpretation of quantum theory due to Sorkin [68, 187, 188], which itself rests on the "principle of preclusion", namely that sets of zero quantum measure do not happen. This is a useful observer independent principle, even in the absence of a complete observer independent quantum interpretation.

5.2 The Partition Function

As we have remarked earlier, causal sets do not admit a natural canonical separation into Cauchy hypersurfaces, even though useful analogues of spatial hypersurfaces do exist. It is therefore much more natural to quantise CST using the path integral or path-sum

$$Z_{\Omega} = \sum_{c \in \Omega} \mu(C), \tag{5.1}$$

where Ω is the **sample space**, and $\mu(C) \in \mathbb{C}$ is a complex valued measure. This is a general prescription, and we must make suitable choices for Ω and $\mu(C)$, depending on the task at hand. A natural choice of sample space is Ω_n, the set of all n-element causal sets where $n \in \mathbb{N}$, but one may also wish to consider the "full" partition function

$$Z_\Omega = \sum_{n=1}^{\infty} Z_{\Omega_n} \tag{5.2}$$

where Ω_n is the set of all n-element causal sets. Each Z_{Ω_n} can, however, be studied independently. Alternatively, we can take $\Omega = \lim_{n\to\infty} \Omega_n$, i.e., the set of all countable n-element causal sets, in which case

$$Z_\Omega = \lim_{n\to\infty} Z_{\Omega_n}. \tag{5.3}$$

As we will see in Sect. 5.2.2 it is also possible to restrict Ω by spacetime or order-theoretic dimension.

More generally, the partition function can be replaced by the double path sum

$$\mathcal{D}_{\Omega_n} = \sum_{C\in\Omega_n} \sum_{C'\in\Omega_n} D(C, C') \tag{5.4}$$

where $\mu(C)$ is replaced by a decoherence functional $D(C, C')$ which satisfies a set of conditions given in Sect. 5.3.3. This is broadly the framework for the quantum sequential growth paradigm of Sect. 5.3. In that case, Ω_n is replaced by the set of past finite countable causal sets Ω_{pf} generated by the sequential growth process, and instead of the fine grained causal set C, one sums over measurable sets.

In this section we consider the conventional, continuum-inspired choice of complex measure

$$\mu(C) \propto e^{\frac{i}{\hbar} S^{(d)}_{BDG}(C)}, \tag{5.5}$$

where $S^{(d)}_{BDG}(C)$ is the d-dimensional discrete Einstein-Hilbert or Benincasa-Dowker-Glaser action on a causal set. A natural choice for d is 4, but one might be justified in using higher dimensional actions in the spirit of Kaluza-Klein theories.

There is also the issue of redundancy or "gauge" in the measure. The natural choice for the sample space is the set Ω of unlabelled causal sets, but often in calculations we have to resort to using the set $\widetilde{\Omega}$ of labelled causal sets. From a physical perspective it is sufficient for the observables to be label-invariant, but it is a matter of choice whether the measure must also be label invariant.

5.2.1 Action Versus Entropy

By replacing the continuum path integral with Eq. (5.1) we expect, of course, to include causal sets that have no continuum approximation. However, we do want the continuum to emerge in the classical limit of the theory. We begin by asking what a typical causal set in Ω_n looks like, in the large n limit. This question was answered

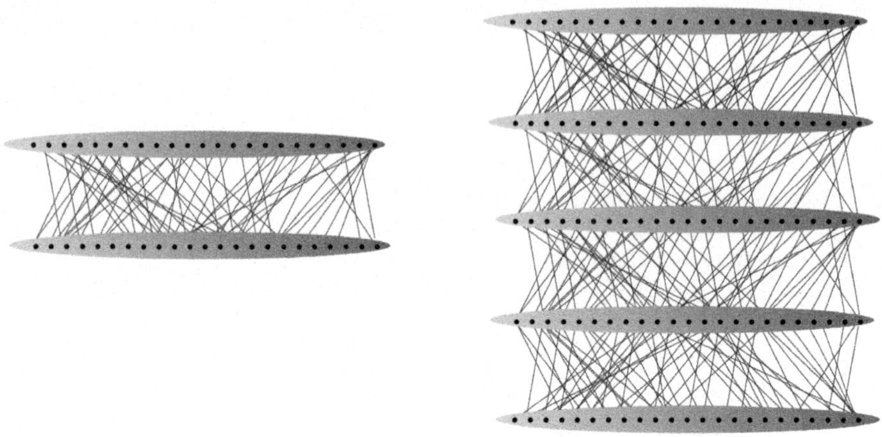

Fig. 5.2 A bilayer poset, and a general K-layered poset

by Kleitmann and Rothschild in [20], where they showed that Ω_n is overwhelmingly dominated by a class of posets we will refer to as **KR posets**, which have no continuum approximation.

Let us begin by defining **K-layered posets** (see Fig. 5.2).

Definition 5.1 A causal set $C \in \Omega_n$ is said to be K-*layered* if it can be partitioned into K disjoint antichains $\mathbb{L}_i$, $C = \mathbb{L}_1 \cup \mathbb{L}_2 \cup \ldots \mathbb{L}_K$, such that:

- If $e \prec e'$ with $e \in \mathbb{L}_k$ and $e' \in \mathbb{L}_{k'}$, then $k < k'$.
- For every $e \in \mathbb{L}_k$ and $e' \in \mathbb{L}_{k'}$ with $k' > k+1$, $e \prec e'$.

Let $\Omega_{\mathrm{K},n}$ denote the set of K-layer posets. For any $c \in \Omega_{\mathrm{K},n}$, let $|\mathbb{L}_i| = \gamma_i n$ where γ_i is the **filling fraction**, with $\sum_{i=1}^{K} \gamma_i = 1$. Since γ_i can be zero for some of the i, $\Omega_{\mathrm{K},n} \subset \Omega_{\mathrm{K}+1,n}$.

Definition 5.2 A KR poset is a 3-layer poset where the layers $\mathbb{L}_i$ satisfy the additional conditions

- $|\mathbb{L}_1|$ and $|\mathbb{L}_3|$ each $\sim \frac{n}{4} + O(\sqrt{n}\log n)$.
- $|\mathbb{L}_2| \sim \frac{n}{2} + O(\log n)$.
- Each $e \in \mathbb{L}_1 \cup \mathbb{L}_3$ is linked to $\sim \frac{n}{4} + O(n^{\frac{7}{8}})$ elements in $\mathbb{L}_2$.
- Each $e \in \mathbb{L}_2$ is linked to $\sim \frac{n}{8} + O(n^{\frac{7}{8}})$ elements of $\mathbb{L}_1$ and similarly, $\mathbb{L}_3$.

Kleitmann and Rothschild showed that the total number of n-element posets grows as $|\Omega_n| \sim 2^{\frac{n^2}{4}+\frac{3n}{2}+o(n)}$ for large n and that the number of KR posets also grows as

$|\Omega_{\rm KR}| \sim 2^{\frac{n^2}{4}+\frac{3n}{2}+o(n)}$. In other words any poset you pick from a bag of n element posets is almost surely a KR poset, when n is large!

The number of layers does not grow with n, which means that the typical large causal set has just three "moments" of time! Any of the tests we perform on KR posets for continuumlike behaviour (for example calculating the Myrheim-Meyer dimension estimator), show that they are non-continuumlike, even under coarse graining.

In [21] Dhar showed that for large n there is a whole hierarchy of sub-dominant K-layer posets for $K \ll n$ which are also not continuumlike. These r**-dominant** posets are parameterised, in addition, by the number of relations rn^2 where $2r$ is the ordering fraction.

Definition 5.3 An n-element $K(r)$-layer poset is r**-dominant** for a fixed ordering fraction $2r$ if (a) the number of links between *any* two consecutive layers is $p(r)$, and $|\widetilde{\Omega}_{{\rm K},n}| = 2^{c(r)n^2+o(n^2)}$, where $c(r), k(r), p(r)$ are certain fixed functions of r with $c(r) \leq \frac{1}{4}$ [21,189] and (b) for every $e \in \mathbb{L}_k$, $e' \in \mathbb{L}_{k+2}$, $e \prec e'$.

The r-dominant posets for $r < \frac{3}{16}$ are

1. $0 < r \leq \frac{1}{8}$, $K(r) = 2$, $c(r) = \frac{1}{4}$, with $\gamma_1 = \gamma_2 = \frac{1}{2}$ which are the symmetric bilayer ($k = 2$-layer) posets.
2. $\frac{1}{8} \leq r \leq \frac{3}{16}$, $K(r) = 3$, $c(r) = \frac{1}{4}$, with $\gamma_1 = \frac{1}{2} - x$, $\gamma_2 = \frac{1}{2}$, $\gamma_3 = x$, $x = \frac{1}{4} - \sqrt{\frac{3}{16} - d}$ which are the KR posets.

For sufficiently large n, the results of [21,189] imply that for $K \ll n$, the dominant contribution comes from KR posets ($K = 3$) with the first weakly sub-leading contribution from bilayer posets ($K = 2$), the next from $K = 4$-layer posets and so on. The K-layer contribution when $K \ll n$ is $|\Omega_{{\rm K},n}| \sim 2^{c(K)n^2+o(n^2)}$, where for $K \geq 4$, $c(K) < \frac{1}{4}$, and is a monotonically decreasing function of K.

It is immediately obvious from the experience we've gained from Chap. 3 that none of these layered posets are continuumlike. Hence Ω_n is dominated, sub-dominated, etc., by causal sets that are non-continuumlike. As we've mentioned before, this entropic dominance of non-continuumlike geometries is shared by other approaches to quantum gravity, for example $d = 2$ Euclidean quantum gravity, with the proliferation of baby universes (branched polymers) [51]. In ordinary free particle quantum mechanics as well, the path integral is entropically dominated by highly non-classical paths, but the action comes to the rescue and restores classicality as $\hbar \to 0$. In this case, there is destructive interference from the rapidly oscillating phases from non-classical paths. One needs similar "miracles" to happen in the quantum path integral, to restore the continuum in the right regime. What we will see is that such a partial miracle does take place for causal sets, for the K-layer posets, despite all odds!

Exercise 5.1 Find the Myrheim-Meyer dimension for $K = 2$ and $K = 3$-layer r-dominant causal sets. Suggest tests for showing that these are not manifoldlike. How does your argument change with K? Use [50] if needed to help construct explicit examples of r-dominant K-layer posets.

The contribution of the class of dominant K-layered causal sets to the partition function can be re-expressed as

$$Z_{K,n} = \sum_{\Omega_{K,n}} e^{\frac{i}{\hbar}S_{BDG}(C)} = \sum_{S_{\min}}^{S_{\max}} \mu(S) e^{\frac{i}{\hbar}S} \tag{5.6}$$

where $S_{\min}$, $S_{\max}$ are the minimum and maximum values of the BDG action for a given $\Omega_{K,n}$, and $\mu(S)$ is the number of causal sets $C \in \Omega_{K,n}$ with $S_{BDG}(C) = S$. Counting the number of such "iso-action" causal sets is of course non-trivial, even when restricted to the set $\Omega_{K,n}$.

Let us begin with the simplest case of $K = 2$ or bilayer posets and the construction of Loomis and Carlip [52]. Here, the only possible intervals N_i are links, i.e., $i = 0$. The BDG action Eq. (3.97) reduces to the **link action**

$$S^{(d)}_{\text{link}}(C) = -\alpha_d \left(\frac{\ell}{\ell_p}\right)^{d-2} (n + \frac{\beta_d}{\alpha_d} N_0), \tag{5.7}$$

where α_d, β_d are given in Eqs. (3.91), (3.92). Thus, the number of iso-action bilayer causal sets simplifies to those with the same number of links [52]. For bilayer orders, the maximum number of possible links is $n^2/4$ and occurs only for the filling fraction $\mathbf{q}_b = (1/2, 1/2)$. Define $N_0 \equiv pn^2 \Rightarrow p \leq \frac{1}{4}$. In the large n limit, the parameter p can be assumed to be continuous and hence [52]

$$Z_2 \approx \text{constant} \times \int_0^{\frac{1}{4}} dp \mu_2(p) \exp(i\omega_d pn^2), \quad \omega_d \equiv -\left(\frac{\ell}{\ell_p}\right)^{d-2} \beta_d, \tag{5.8}$$

where $\mu_2(p)$ can be estimated to leading order as follows. For a given p, let us first fix the filling fraction to $(\gamma, 1-\gamma)$. The maximum possible number of links is $\gamma(1-\gamma)n^2$ and hence the number of labelled n-element causal sets $\mathcal{P}_{\gamma,p,n}$ with filling fraction γ and pn^2 links is

$$|\mathcal{P}_{\gamma,p,n}| = \binom{n}{\gamma} \binom{\gamma(1-\gamma)n^2}{pn^2}, \tag{5.9}$$

where the first factor accounts for the possible number of labelled bi-partitions for a given γ. Thus, the largest contribution comes from $\gamma_{\max} = \frac{1}{2}$. Using Stirling's approximation, to leading order

$$\ln |\mathcal{P}_{\gamma,p,n}| = \Big(\gamma(1-\gamma)\ln(\gamma(1-\gamma)) - p\ln p - (\gamma(1-\gamma) - p)\ln(\gamma(1-\gamma) - p)\Big)n^2 + o(n^2). \tag{5.10}$$

For $\gamma = \frac{1}{2}$

$$\ln |\mathcal{P}_{\frac{1}{2},p,n}| = \frac{1}{4}h(4p)n^2 + o(n^2), \tag{5.11}$$

where

$$h(\tilde{p}) = -\tilde{p}\ln\tilde{p} - (1-\tilde{p})\ln(1-\tilde{p}), \qquad 0 < \tilde{p} < 1, \tag{5.12}$$

is **Dhar's entropy function** [21].

Exercise 5.2 Let $n = 4$, and $p = \frac{1}{4^2}$. Show that the sets $\mathcal{P}_{\gamma,\frac{1}{4^2},4}$ for different values of γ are not necessarily disjoint.

In [52] this expression was used to bound $\mu_2(p)$, the number of bilayer posets with pn^2 links, by noting that as one varies the set $\mathcal{P}_{\gamma,p,n}$ over γ for fixed p, one can generate multiple copies of the same causal set. For example, if $pn^2 = 1$, i.e., there is only a single link, then there can be multiple γ's that give you the same causal set. As a result,

$$\mu_2(p) \leq \sum_{\gamma=0}^{\frac{1}{2}} |\mathcal{P}_{\gamma,p,n}| \approx 2^{\frac{1}{4}h(4p)n^2+o(n^2)}, \tag{5.13}$$

since the latter is dominated by the largest contribution from $\gamma_{\max}$ in its exponent. Similarly, there is a lower bound

$$|\mathcal{P}_{\frac{1}{2},p,n}| \leq \mu_2(p), \tag{5.14}$$

which, when combined with Eq. (5.13) implies that to leading order

$$\mu_2(p) = \frac{1}{4}h(4p)n^2 + o(n^2). \tag{5.15}$$

Thus,

$$Z_2 \approx c\int_0^{\frac{1}{4}} dp\, \exp(i\omega_d pn^2 + \frac{1}{4}h(4p)n^2 + o(n^2)). \tag{5.16}$$

This integral can be evaluated to leading order using the method of steepest descent about the saddle point at

$$p_0 = \frac{e^{i\alpha_0}}{4\cos\omega_d}. \tag{5.17}$$

To get a suppression for Z_2, the parameter ω_d must satisfy

$$\tan\left[\frac{1}{2}\beta_d\left(\frac{\ell}{\ell_p}\right)^{d-2}\right] > \left(\frac{27}{4}e^{-1/2} - 1\right)^{1/2}. \tag{5.18}$$

where we have used the fact that the dimension dependent constant $\omega_d < 0$ in the BDG action, in any dimension. This condition is satisfied whenever $\ell > \ell_0$, where $1.13 < (\ell_0/\ell_p) < 2.33$, which is a very conservative constraint on the discreteness scale. For $d = 4$ in particular, $\ell_0 \approx 1.136\ell_p$.

This result of Loomis and Carlip is a rather remarkable analytical, non-perturbative result in quantum gravity, since it simultaneously gives a *lower* bound on the discreteness scale of order the Planck scale. In other words, it states that if CST were to have a classical, continuum approximation in the large n limit, then the discreteness scale must be at least of order the Planck scale, in any dimension! Almost nothing "extra" has gone in—no symmetries and no dimensional reduction.

As we will see, this lower bound on the discreteness scale continues to be true even when one includes all K-layer posets include the most dominant, the KR posets [53–55]!

The main technical difficulty in extending the bilayer result to $K > 2$, is that the BDG action is no longer the link action but includes contributions from higher element intervals. Thus, one would have to estimate the number of iso-action causal K-layer posets.

Consider the KR poset. Its connectivity properties gives us a hint that N_0 dominates the other N_i at least for the small non-zero i that appears in the $d = 4$ BDG action.

Exercise 5.3 Calculate the N_i for the 16 element KR poset in Fig. A.7. Simulate a random KR poset for $n \sim 50$ and plot the N_i as a function of i.

One can prove this analytically [54]. A 3-chain $e_1 \prec e_2 \prec e_3$ which is *also* a 1-interval in a KR poset can only occur if $e_1 \in \mathbb{L}_1, e_2 \in \mathbb{L}_2, e_3 \in \mathbb{L}_3$ and there exists no other $e \in \mathbb{L}_2$ such that $e_1 \prec e \prec e_2$. If p is the probability for a link between $\mathbb{L}_1$ and $\mathbb{L}_2$ and q that between $\mathbb{L}_2$ and $\mathbb{L}_3$, then the probability for a pair e_1, e_3 to form a 1-interval is

$$(\gamma_2 n)\, pq\, (1 - pq)^{\gamma_2 n - 1}, \tag{5.19}$$

where the first factor comes from the number of possibles choices for e_2. For the pair e_1, e_3 to form an i-element interval there must be exactly i-elements of $\mathbb{L}_2$ such that $e_1 \prec e \prec e_3$, for which the probability is

$$\binom{\gamma_2 n}{i} pq^i (1 - pq)^{\gamma_2 n - i}. \tag{5.20}$$

Using simple large n approximations, it can be shown that the probability for N_i to be of order $N_0 \sim n^2$, is exponentially small [54]. Hence to leading order, the link action for KR posets *is* the BDG action. A more involved argument in [55] shows that this is also true for all K-layer orders when $K \ll n$.

In other words, to leading order we can get by using just the link action when evaluating the contribution of the K-layer orders to the partition function. In [53] the arguments of [52] were generalised to show that the link action suppresses all the K-layer orders including the KR orders. The saddle point calculation to leading order is the same as that for the bilayer orders and hence the lower bound on the discreteness scale remains unchanged.

As we've emphasised before, these results give us an important non-perturbative result in CST and suggests that the BDG action may be sufficient to ensure that continuumlike causal sets emerge in the semi-classical limit of the theory. For such causal sets, the link action is not likely to be sufficient and one would need to evaluate the number of iso-action causal sets in a given class and to find their dominant characteristics, as was done for the K-layer posets. This is a daunting task to handle analytically, and is compounded by the fact that continuumlike causal sets are themselves hard to characterise. One will doubtless need to supplement any such analytic calculation with numerical simulations.

5.2.2 Lorentzian Statistical Geometry

One of the main objectives in defining a partition function is to calculate the expectation value $\langle \mathbf{O} \rangle$ of physically interesting observables O, where

$$\langle \mathbf{O} \rangle = \frac{1}{Z} \sum_{c \in \Omega} O(c) e^{\frac{i}{\hbar} S_{BDG}(c)} . \tag{5.21}$$

As we have seen, covariant observables in CST are order invariants. In Chap. 3 we found several order invariants, some of which are geometrically significant in the continuum approximation, and others that do not individually have a continuum interpretation.

The generalisation of Eq. (5.6) to Eq. (5.21) would involve not only counting iso-action causal sets, but also those with the same value of O, which is an analytically daunting task, not least because the sample space grows super-exponentially with n. For now, we will try a different route, more in common with other path integral approaches to quantum gravity.

Let us for the moment assume that the action is itself zero. This gives us a uniform probability measure over Ω_n which can be sampled using suitable Markov Chain Monte Carlo techniques [56,58,60,61,184]. In order to construct the Markov chain in Ω_n we need to define a move that converges to the desired "typical" or equilibrium distribution. If the move satisfies ergodicity and detailed balance, then then we are assured of this [190]. Ergodicity means that one can go from any $C \in \Omega_n$ to any other $C' \in \Omega_n$ via a sequence of these moves. A move from $C \to C'$ satisfies detailed balance if

$$\Pr(C) \Pr(C \to C') = \Pr(C') \Pr(C' \to C) , \tag{5.22}$$

where $\Pr(C)$ is the probability to be in state C, and $\Pr(C \to C')$ the transition probability from $C \to C'$. For a uniform distribution, this simplifies to

$$\Pr(C \to C') = \Pr(C' \to C) . \tag{5.23}$$

In a generic graph, a natural move is one where an edge and possibly a vertex is removed or added, to obtain a new graph. Since we have fixed n for now, we do not wish to change the number of vertices, and hence removing an edge in a causal set can mean either removing a relation or removing a link. However, simply adding or removing a relation or a link is not sufficient to attain detailed balance in a causal set since it is a transitively closed graph.

Let us select a pair of elements $e, e' \in C$, uniformly at random. Either $(a)\, e \prec e'$, $(b)\, e' \prec e$ or $(c)\, e \parallel e'$. For (a), $(i)\, e \prec\!\!\ast\, e'$ or $(ii)\, e \not\prec\!\!\ast\, e'$. If the latter, then removing the relation is not enough, since there are elements e'' such that $e \prec e'' \prec e'$. Removing the relation implies that transitivity is lost, and restoring transitivity will restore the same relation and hence make the move trivial. One needs to not only remove the relation $e \prec e'$, but also all the relations from e, e' to the elements in $I(e, e')$, as well as relations between $e'' \in I(e, e')$ and those in $\mathrm{Past}[e] \cup \mathrm{Fut}[e']$ that are induced by transitivity *via* e or e'. This makes for a very messy move! Thus option (ii) gives rise to a situation that is computationally too messy and we should therefore go back to picking another pair uniformly at random.

If (i) is true, i.e., e and e' are linked then we can remove the link. At the same time we have to either

- (R) leave all other relations as is. This will mean that new links are formed in the process, since for any linked-path $e_1 \prec\!\!\ast\, e \prec\!\!\ast\, e'$ in C the removal of the link $e \prec\!\!\ast\, e'$ *can* lead to the new link $e_1 \prec\!\!\ast\, e'$ unless there is an $e'' \in C, e'' \neq e$ such that $e_1 \prec e'' \prec e'$.
  OR
- (L) remove relations between elements of $\mathrm{IPast}(e)$ and elements of $\mathrm{IFut}(e')$ unless they are implied by transitivity via an element $e'' \parallel e, e'$. This process does not create new links.

The former is called the **relation move**, with $C \to C_R$, since the number of relations goes down by exactly one but not the number of links, and the latter the **link move** with $C \to C_L$ since the number of links goes down by exactly one, but the number of relations can be reduced.

One has to now be able to reverse these moves, which means we start with $e \parallel e'$. Let us add in a link $e \prec\!\!\ast\, e'$. This process can lead to the further removal of a link (but not a relation) that could have existed between $e_1 \in \mathrm{IPast}[e]$ and $e' \in \mathrm{IFut}[e']$, but also adds in several relations because of transitivity. Thus, unless we make further demands on e, e' we will find that the move is not reversible either to (R) or to (L).

- For the move (R) we notice that after the link is removed, $\mathrm{Past}(e) \subseteq \mathrm{Past}(e')$ and $\mathrm{Fut}(e') \subseteq \mathrm{Fut}(e)$. This is the requirement that we should place on $e \parallel e'$ in order to reverse the (R) move. This way, the number of relations goes up by just one, thus reversing the (R) move. We call such pairs "critical".
- For the (L) move, we notice that after the link is removed, no new links are created between elements in $\mathrm{IPast}(e)$ and $\mathrm{IFut}(e')$. Therefore we should ensure that when we add in the link to $e \parallel e'$, there are no links that are removed, which means that

there should be no links between elements in IPast(e) and IFut(e'). We call such pairs "suitable".

The question we have to ask next is whether these two moves are ergodic, i.e., do they converge to the same state for any initial configuration? In the simulations [58], several different initial states were considered, including the chain, the antichain and the KR poset. Tracking the expectation value of various observables it was shown that there is rapid convergence to the same final state.

Next, one must check if R and L satisfy detailed balance. There are $\binom{n}{2}$ pairs e, e' and hence the probability of choosing a particular pair is $\frac{2}{n(n-1)}$. If there is a link, then the link is removed. If it is an R move $C \to C_R$, e, e' become a critical pair, and if it is an L move $C \to C_L$ they are a suitable pair. If this same pair is again chosen in $C_{R,L}$ with probability $\frac{2}{n(n-1)}$, then it satisfies criticality and suitability, respectively, and the move means that a link is added from e to e' to get C. Thus, we see that both R and L moves satisfy detailed balance.

On the computer, causal sets are stored either as causal matrices or link matrices, and therefore come labelled. We perform our moves on $\widetilde{C} \in \widetilde{\Omega}_n$ the set of n-element labelled causal sets. For causal sets that are naturally labelled, the causal or link matrix is upper triangular, and we can make our search for linked pairs more efficient by requiring that $l(e) < l(e')$ where $l : C \to \mathbb{N}$ is a **natural labelling**. However, physical, covariant observables are label invariant, so we need to be able to remove this "gauge" from our calculations. While it is possible to check if two labelled causal sets correspond to the same unlabelled causal set for very small n, it becomes computationally expensive to do this as n becomes larger. Thus, "gauging out" this extra freedom is not practical.

Because the R or L moves change the number of relations and/or links, if $\widetilde{C} \to \widetilde{C}'$, then $C = [\widetilde{C}] \not\sim C' = [\widetilde{C}']$, namely they are not related via relabelling. If $\widetilde{\Omega}_n$ is the set of all labelled causal sets, then one can remove the gauge, *upto automorphisms* by dividing by $n!$. As n increases, the number of causal sets with non-trivial automorphisms is exponentially suppressed [191,192], and hence $n!$ is a good enough measure of the gauge group.

The choice of R or L moves allows us to move in $\widetilde{\Omega}$ and to find "typical" or equilibrium configurations. Once this point of equilibration is obtained, it should not matter what initial causal set C one starts from. In order to assess the approach to equilibrium, or "thermalisation" one has to track the behaviour of a suite of order invariants (which are the analogues of order parameters in the Ising system.) For the R and L moves, which involves a pair of elements at every step, it is important to sweep through all pairs at least once before taking a measurement of these order invariants. Thus we define a **sweep** as a set of $n(n-1)/2 \sim n^2/2$ steps. Once thermalisation is achieved, the autocorrelation time must be found; only measurements separated by the autocorrelation time can be considered truly independent.

In [58] the R and L moves were implemented upto $n = 84$ on the set of naturally labelled causal sets $\overline{\Omega}$. For the uniform measure, since asymptotically the typical causal set is a KR poset, we can ask whether these simulations generate KR orders even at these small values of n. The KR orders are characterised by their ordering fraction 3/8, their linking fraction 1/2, the filling fraction $|\mathbb{L}_i|/n = \gamma_i$ for its layers as well as a height $h = 3$. Tracking these order invariants shows that there is an onset of KR behaviour as we reach $n \sim 84$. The R and L moves, however, are quite inefficient since there is a high rejection rate once a pair is selected at random. This limits the efficiency of the code as one goes to larger n-values. Importantly, the nonlocality of the causal set means that seemingly local moves like the R and L moves can have a global impact in the causal set, and can lead to drastic changes in the order invariants. This means that it is also not possible to parallelise the code, at least naively, since one cannot simply "add up" local effects of moves that take place in different parts of the causal set, as in the Ising model.

Exercise 5.4 Construct Markov chains of length 10 for the L type and R type move starting from the 3-element antichain $\mathrm{a}_3 \equiv \cdots$, the 3-element chain $\mathrm{c}_3 \equiv \vdots$, and the "$\Lambda$" causal set $\Lambda_3 \equiv \Lambda$. Track the ordering fraction r along the Markov chain. You may also want to try your hand at writing a code for small n (it need not be efficient!) Does it thermalise with respect to r?

The simulations of [58] show that it is possible to sample the space of causal sets Ω_n using standard MCMC techniques when the measure is uniform. However, for the quantum partition function, the measure is complex, and such "real" sampling techniques are not adequate. Moreover, the expectation value $\langle \mathbf{O} \rangle$ of an order invariant O given in Eq. (5.21) need not be real, and leads to interpretational ambiguities.

One route is to take our cue from Quantum Field Theory, where the quantum partition function is analytically continued to a real statistical partition function. We do not have the ability to perform a Wick rotation $t \to it$ in a causal set because of the lack of global time coordinates, and hence one must find alternative routes.

In [56] an analytically continued partition function was obtained by introducing a new inverse temperature-like parameter β into the partition function

$$\widetilde{Z}_\beta = \sum_{c \in \Omega_n} e^{i \frac{\beta}{\hbar} S(c)}. \tag{5.24}$$

The analytic continuation $\beta \to i\beta$ makes the measure real, without changing the sample space. We define

$$Z_\beta \equiv \widetilde{Z}_{i\beta} = \sum_{c \in \Omega_n} e^{-\frac{\beta}{\hbar} S(c)}. \tag{5.25}$$

Thus, there is no "Euclideanisation" of the causal set analogous to Wick rotated Minkowski spacetime, and Eq. (5.25) gives us a theory of **Lorentzian statistical geometry**.

Because of the difficulty in the MCMC simulations for Ω_n, the first fully non-perturbative calculation of Z_β was done for a dimensionally restricted sample space $\Omega_{2d}(n) \subset \Omega_n$ of "2d orders". These are n-element orders that are constructed from the intersection of two linear or total orders. Consider two linear orders $U = (u_1, u_2, \ldots u_n)$ and $V = (v_1, v_2, \ldots v_n)$, where $u_i, v_i \in S \equiv \{1, \ldots, n\}$, such that for $i \neq j, u_i \neq u_j$ and $v_i \neq v_j$. U and V are totally ordered since for every pair $i \neq j$ either $u_i < u_j$ or $u_j < u_i$, and similarly for V. A **2d order** $C = U \cap V$ has elements $e_i = (u_i, v_i)$, which obey the order relation $e_i \prec e_j$ iff $u_i < u_j$ and $v_i < v_j$. Figure 5.3 shows examples of $2d$-orders. It is trivial to generalise this construction to obtain n-d orders as an intersection of n linear orders.

An immediate question that comes to mind is whether this order theoretic dimension has anything to do with the dimension of spacetime. In $d = 2$ the answer is in the affirmative. Consider a causal diamond $\mathbb{D}^2$ in $d = 2$ Minkowski spacetime with light-cone coordinates $u, v \in [-L, L]$. Construct a light cone lattice with grid spacing $\frac{2L}{n+1}$ so that on the grid, u, v take values $S = \{L(-1 + \frac{2}{n+1}), L(-1 + \frac{4}{n+1}), \ldots, L(-1 + \frac{2n}{n+1})\}$. Define linear orders U and V with $u_i, v_i \in S$. The events $e_i \equiv (u_i, v_i)$ in $\mathbb{D}^2$ are clearly ordered according to the intersection $U \cap V$ and hence the associated causal set is an n element 2-d order. Thus, any n-element causal set that embeds into $\mathbb{D}^2$, not necessarily faithfully, is a 2-d order.

Exercise 5.5 Which of the $n = 4$ element posets in Appendix A are 2-d orders? Hint: use the set of $n = 4$ integers to define U and V. Show that while the $n = 4$ element crown poset in Fig. A.5 is a 2-d order, this is not true of the 6-element crown poset.

Winkler, and independently El-Zahar and Sauer [193, 194], showed that a typical 2-d order is a **random order**, i.e., one in which the elements in the total orders U and V are chosen uniformly at random. In [56] it was realised that a random 2-d order is continuumlike and approximated by the causal diamond $\mathbb{D}^2 \subset \mathbb{M}^2$. This is a strong departure from Ω_n which is overwhelmingly dominated by the non-continuumlike causal sets, in particular the KR posets.

Exercise 5.6 A k-dimensional order is made up of an intersection of k linear orders $U_1 \cap U_2 \ldots U_k$. Do these orders embed into a causal diamond in $\mathbb{M}^k$? Use general arguments to show that for $k > 2$ the typical random k-dimensional order is non-continuumlike.

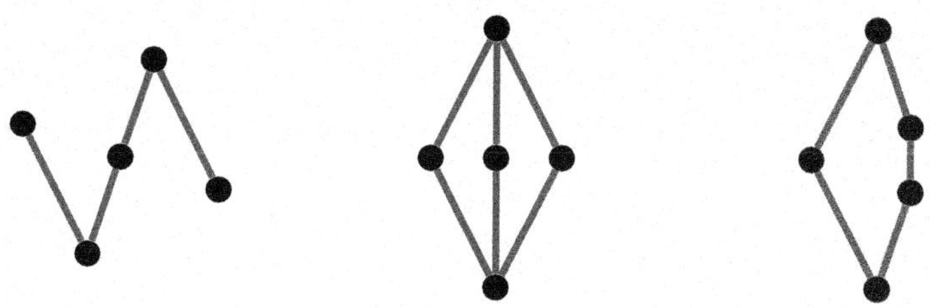

Fig. 5.3 Examples of 2d orders

Not every causal set in $\Omega_{2d}(n)$ embeds faithfully into $\mathbb{D}^2$. Hence a restriction to $\Omega_{2d}(n)$ can be considered to be a fair representation of the history space of topologically trivial $d = 2$ causal set theory. We define the analytically continued partition function

$$Z_\beta^n = \sum_{\Omega_{2d}(n)} \exp(-\beta S_{BDG}^{(2)}(C)), \tag{5.26}$$

where we have chosen $S_{BDG}^{(2)}(C)$ to be the $d = 2$ BDG action.

We could use the moves R and L described above to traverse $\Omega_{2d}(n)$, but the nature of the 2-d order allows for a much more efficient combinatorial move. On the computer we take $S = \{1, \ldots n\}$ which corresponds to a latticisation of $\mathbb{D}^2$ with $u, v \in [0, n+1]$. We define the **exchange** (E) move. For any $C = U \cap V$

- Pick either U or V at random.
  - If U is picked
    - Pick a pair (i, j) at random from S, with $i \neq j$.
    - Swap u_i and u_j in U to obtain U'.
    - Then $C' \equiv U' \cap V$
  - If V is picked,
    - Pick a pair (i, j) at random from S, with $i \neq j$.
    - Swap v_i and v_j in V to obtain V'.
    - Then $C' \equiv U \cap V'$

Figure 5.4 shows an example of this move starting from the n-chain and the n-antichain.

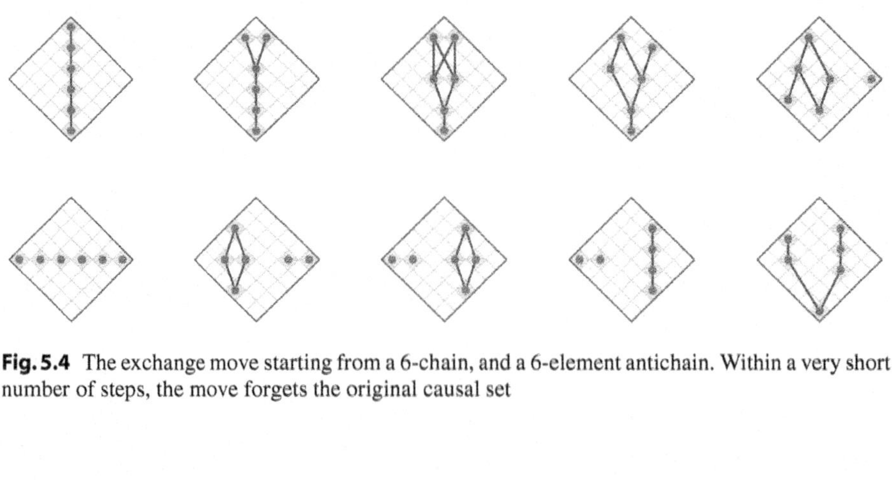

Fig. 5.4 The exchange move starting from a 6-chain, and a 6-element antichain. Within a very short number of steps, the move forgets the original causal set

This move is clearly ergodic and satisfies detailed balance. In order to improve efficiency, we further use the Metropolis-Hastings algorithm to move the MC forward, by accepting a move $C \to C'$ if the difference in the action $S^{(2)}_{\rm BD}(C') - S^{(2)}_{\rm BD}(C)$ is negative, and rejecting a move only if $\exp^{-\beta(S^{(2)}_{\rm BD}(C')-S^{(2)}_{\rm BD}(C))} < a$, where a is a random number in $[0, 1)$.

One can also vary over different mesoscale values $\epsilon \in (0, 1]$ and use the more general $d = 2$ action (Eq. (3.101))

$$\frac{1}{\hbar} S^{(2)}_{\rm BD}(c, \epsilon) = 4\epsilon\left(n - 2\epsilon \sum_{i=0}^{n-2} N_i f(i, \epsilon)\right), \tag{5.27}$$

where $\epsilon = l_p^2/l_k^2$ and $f(i, \epsilon) = (1-\epsilon)^i(1 - 2\epsilon i(1-\epsilon)^{-1} + \epsilon^2 i(i-1)(1-\epsilon)^{-2}/2)$.

Extensive numerical results have been carried out for this system [56,59,184]. As in the case of the uniform measure over Ω_n, a single sweep is taken to be $n(n-2)/2$ steps since, as with the R and L moves, the E move involves choosing a distinct pair of elements (e_i, e_j) over n elements. Different order invariants can be calculated at every sweep, including the ordering fraction r, the number of i-element intervals N_i, the number of maximum and minimum elements, and the height.

The process is said to be ergodic if it explores all of the possible states. This is of course not possible in finite computation time. Another way to check this is to see the dependence on the initial state. The Markov Chain was started from eight different types of initial 2d orders, including the random 2d order, the chain, and the antichain. As can be seen in Fig. 5.5, all these initial states rapidly converge to a single value of the order invariant. Even though thermalisation has not been reached, the information about the inital state is lost within the first few sweeps.

Once thermalisation sets in for a large enough number of sweeps, the autocorrelation times τ_O for each order invariant can be calculated. Picking values at intervals much larger than the autocorrelation time, the "thermalised" expectation values for the various order invariants O is then found for each value of β and ϵ (see Fig. 5.6).

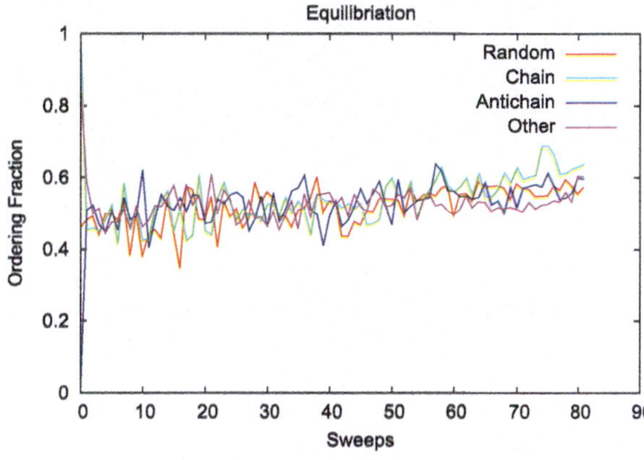

Fig. 5.5 An example of the ergodicity in the MCMC process for the 2d order using the BDG action as the order invariant. Within a few sweeps there is no trace of the initial causal set

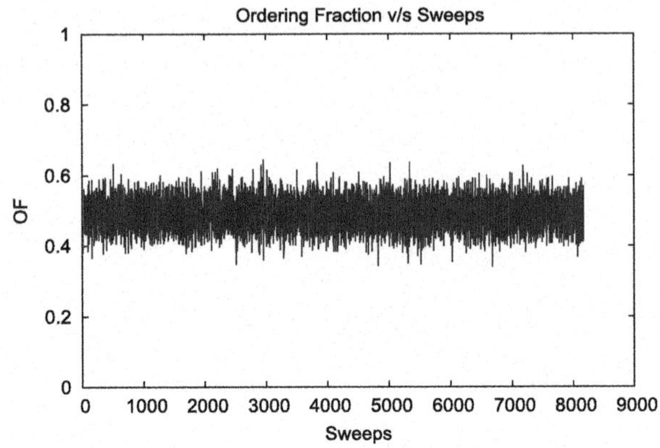

Fig. 5.6 A thermalisation plot for the ordering fraction versus the number of sweeps for $N = 50$, $\epsilon = 0.1$ and $\beta = 0.8$. Thermalisation occurs within the first couple of sweeps

The behaviour of $\langle \mathbf{O} \rangle$ with β shows a marked and rapid change at a critical value β_c, signalling a phase transition. On closer examination, we see two distinct phases on either side of β_c. For $\beta < \beta_c$ the expectation value of the order invariants matches that of causal sets approximated by $\mathbb{D}^2$, while for $\beta > \beta_c$, the order invariants are distinctly non-continuumlike, as shown in Figs. 5.7 and 5.8.

Varying over both ϵ and β, we get the phase diagram shown in Fig. 5.9 where the critical values $\beta_c(\epsilon)$ clearly separate the two phases. In [59] an extensive finite size scaling analysis was done for this system and it was found that a rescaling of β_c leads to a collapse of the data as shown in Fig. 5.10. Moreover, a careful analysis of the

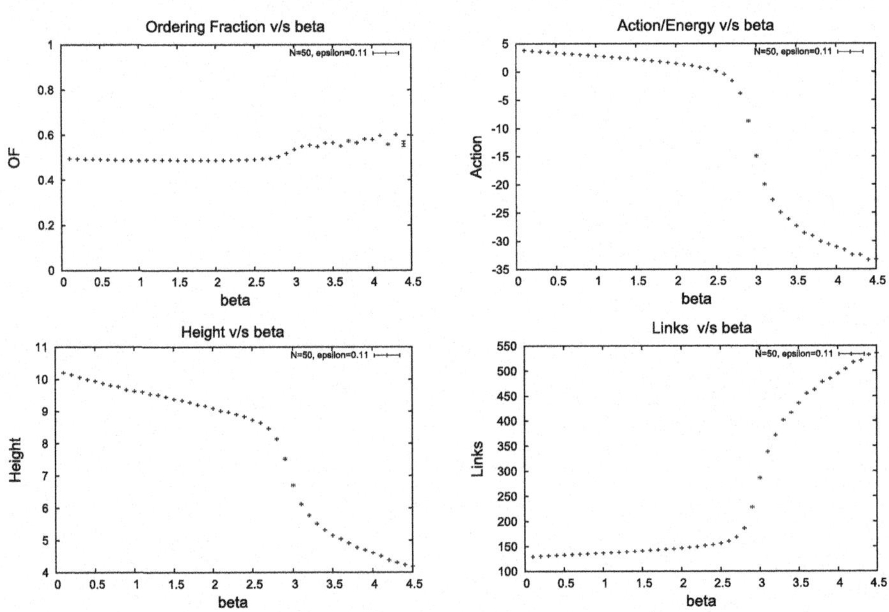

Fig. 5.7 Expectation values of various covariant observables or order invariants are plotted as a function of β for $N = 50$, $\epsilon = 0.11$. The error bars are very small and though they are included, can barely be seen. The fairly sharp change in the values is an indication of a phase transition

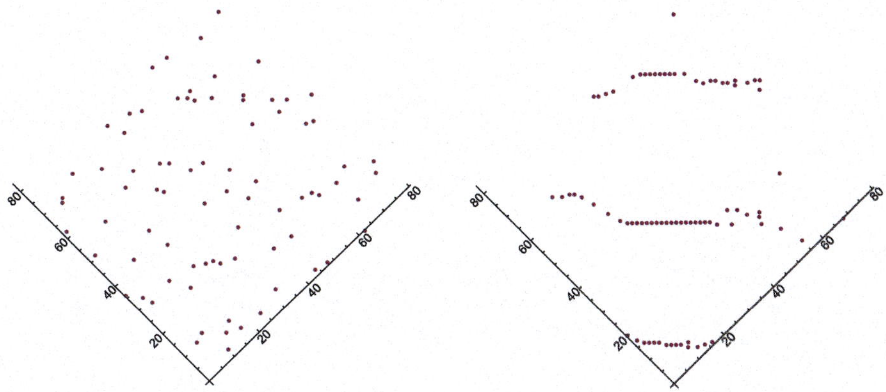

Fig. 5.8 Results from simulations of the MCMC for 2d orders for $n = 80$. At small β values the typical causal set is continuumlike, while at large β values it has a layered structure

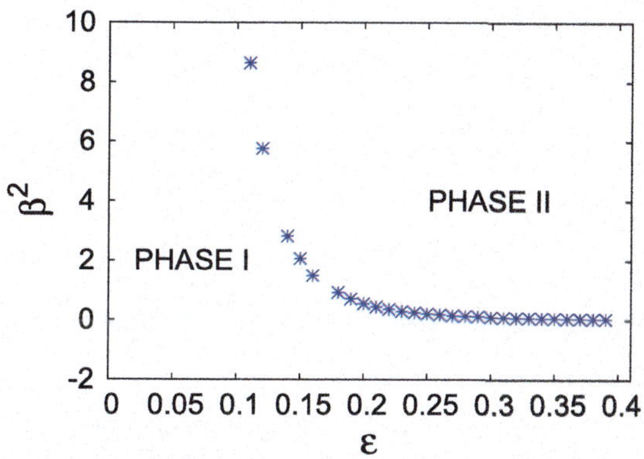

Fig. 5.9 The critical line β_c^2 is a function of ϵ separating the continuumlike phase in $\beta^2 < \beta_c^2$ and the non-continuumlike phase $\beta^2 > \beta_c^2$

distribution of the causal sets in the ensemble around β_c shows a double Gaussian, which is characteristic of a first order phase transition (see reference [59] for more details.)

Since the 2d orders are not only dimension restrictions but are also topologically trivial, the next step is to look for generalisations. As the exercise above shows, the k-d orders are not the right higher dimensional generalisations and one has to resort to a different way of generating the sample space. We take our cue from the fact that all 2d orders embed into the lightcone latticisation of $\mathbb{D}^2$, although not faithfully. This gives us $m = n^2$ sites with n occupations, with the further constraints that the u_i and v_i values can never be repeated. The idea is to generalise this to topologies and geometries of interest to us.

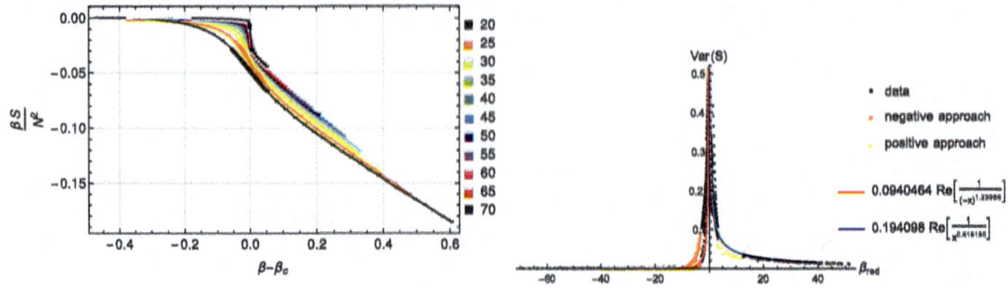

Fig. 5.10 A rescaling of $\beta\langle S(C)\rangle/n^2$ collapses the action data for different n values. The figure on the left shows the collapse for the action for $\epsilon = 0.35$, while the figure on the right shows the collapse in the specific heat, or $var(S)$ for $\epsilon = 0.23$

We begin with a (t, θ) latticisation of the topologically nontrivial $d = 2$ flat cylinder spacetime $\mathbb{S}^1 \times I$

$$ds^2 = -dt^2 + d\theta^2, \tag{5.28}$$

with $m = T \times \Theta$ number of lattice sites which are occupied by n causal set elements. This gives us a $T \times \Theta$ **occupation matrix** of the type

$$\mathbf{M} = \begin{pmatrix} 0\,0 \cdots & 1 & & 0 & 1 \\ 1\,0\;\;0 & 1 & \cdots & 0 & 1 \\ 0\,1 \cdots & 0 & 0 & 1 & 0 \\ 0\,0\;\;0 & \cdots & 0 & 1 & 0 \\ \vdots\;\vdots\;\;\vdots & \vdots & \vdots & \vdots & \vdots \\ 1\,0\;\;0 & 0 & 0 & \cdots & 1 \end{pmatrix} \tag{5.29}$$

The causal matrix is then easy to extract, using the metric Eq. (5.28). This gives us a sample space of labelled causal sets Ω_{cyl}. Similarly, the latticisation of $d = 3$ flat spacetime with toroidal spatial slices $T^2 \times I$

$$ds^2 = -dt^2 + d\theta^2 + d\phi^2, \tag{5.30}$$

yields a sample space of causal sets $\Omega_{T^2 \times I}$, using a similar 3d occupation matrix. The Markov chain on these sample spaces can be generated using a natural **lattice-gas move** on the occupation matrix: an occupied site is emptied and an unoccupied site is filled, as shown in Fig. 5.11.

These systems were extensively studied in [61] using MCMC techniques. As with the 2d orders, the MCMC simulations using this move thermalise rapidly, allowing the accurate calculation of the expectation values of order invariants as a function of β and the mesoscale ϵ. Unlike the case of 2d orders, we have one more free parameter, namely the number of sites m. Simulations show that as $m >> n$, the dependence on m drops out.

Like the 2d orders, these cases also exhibit a first order phase transition which separate a continuumlike phase from a layered phase as shown in Figs. 5.12, 5.13. In retrospect, one might not be surprised at the appearance of the continuum phase,

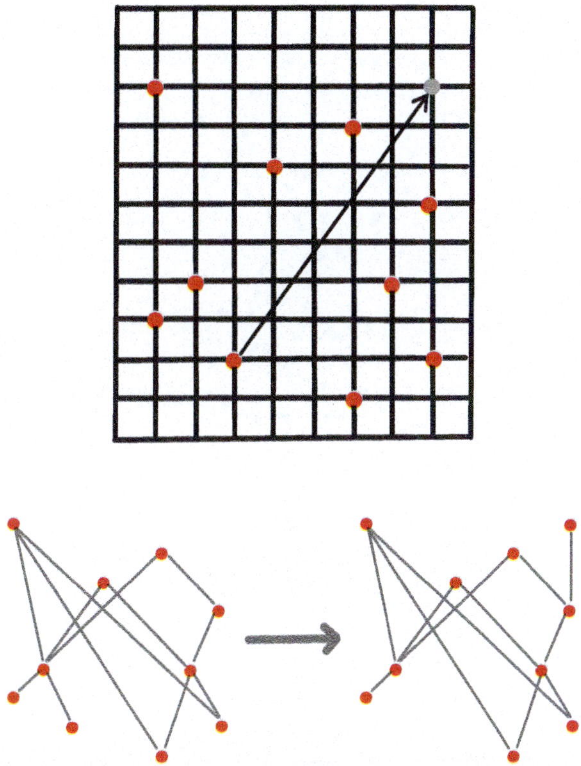

Fig. 5.11 An example of a lattice gas move with a filled site being emptied to an unfilled site. This leads to change in the embedded causal set, as shown on the right

since it mimics Winkler and El-Zahar and Sauer's 2d result [193, 194] that the typical order is a random order and hence a Poisson sprinkling into the lattice. However, the existence of a clear phase transition of a similar character could not perhaps have been predicted on purely analytic grounds. Generalising to arbitrary dimensions is possible in principle but simulations are challenging.

For those looking for a continuum limit, a first order phase transition is not sufficiently good news. In [60] the 2d orders were combined with an Ising spin system, to see if the addition of matter can modify the phase structure. However, the results do not suggest that this is the case.

It should be remembered that the continuum limit is not physically relevant to causal set theory, but only the continuum approximation. Therefore it is more fruitful to seek an interpretion of these results from a more physical, cosmological perspective. To begin with, we need to ask how the analytically continued path integral Eq. (5.25) itself may be understood in the full quantum theory of causal sets. In Lattice Gauge Theories as also in CDTs, one appeals to reconstruction theorems to show that the Euclidean, statistical theory can be reinterpreted as a Lorentzian, quantum theory by satisfying the condition of reflection positivity [3, 195]. However, as formulated, showing this in causal set theory is nontrivial, especially since it is not amenable to Hamiltonian methods.

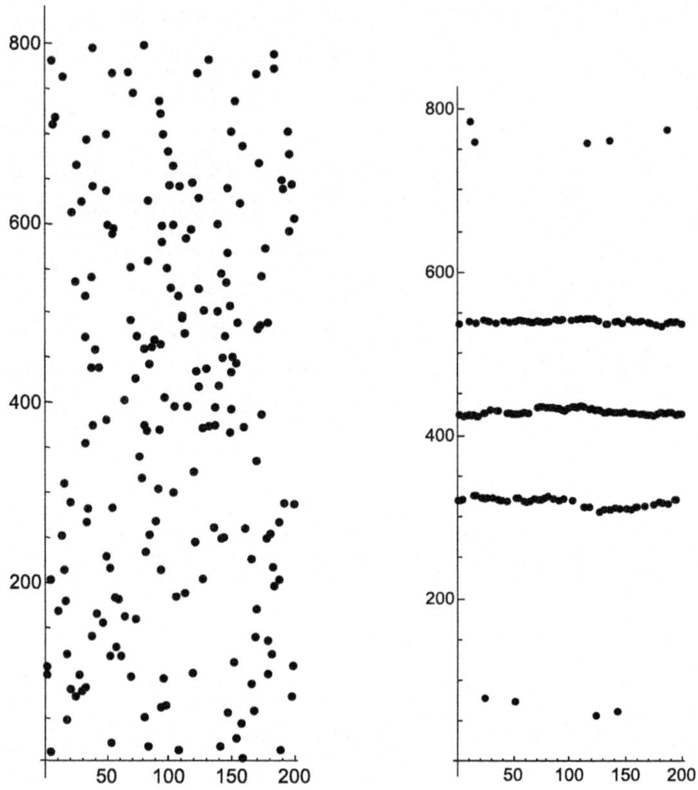

Fig. 5.12 Results from simulations of the MCMC for the 2d cylinder spacetime $\mathbb{S}^1 \times I$ with $n = 200$ elements. The low β or "hot" phase is continuumlike, while the high β, or "cold" phase is highly layered and hence non-continuumlike

However, Eq. (5.25) may be interpreted in the Hartle-Hawking picture as a ground state wave function [57]. In the continuum, this is the ground state over $(d-1)$ geometries (Σ, h) and is given by the Euclidean functional integral over (M, g)

$$\Psi_0(h_{ab}, \Sigma) = A \sum_M \int dg^E e^{-I_E(g)}, \tag{5.31}$$

where $\partial M = \Sigma$, $g|_\Sigma = h$, $I_E(g)$ is the Euclidean Einstein action and A is a normalisation constant [57]. $\Psi_0(h_{ab}, \Sigma)$ is thus a functional over all closed 3-geometries and is the initial state of the universe from which further evolution of the wave function can be (uniquely) determined. There is only one "final" boundary ∂M in Eq. (5.31), and no "initial" boundary.

While a final spatial boundary in a causal set may not be strictly Cauchy to the future, this allows us to assign a more physical meaning to the Lorentzian statistical path sum. Instead of a Euclidean geometry, the no-boundary condition for a causal set can be viewed as the requirement that there is a single element to the past of

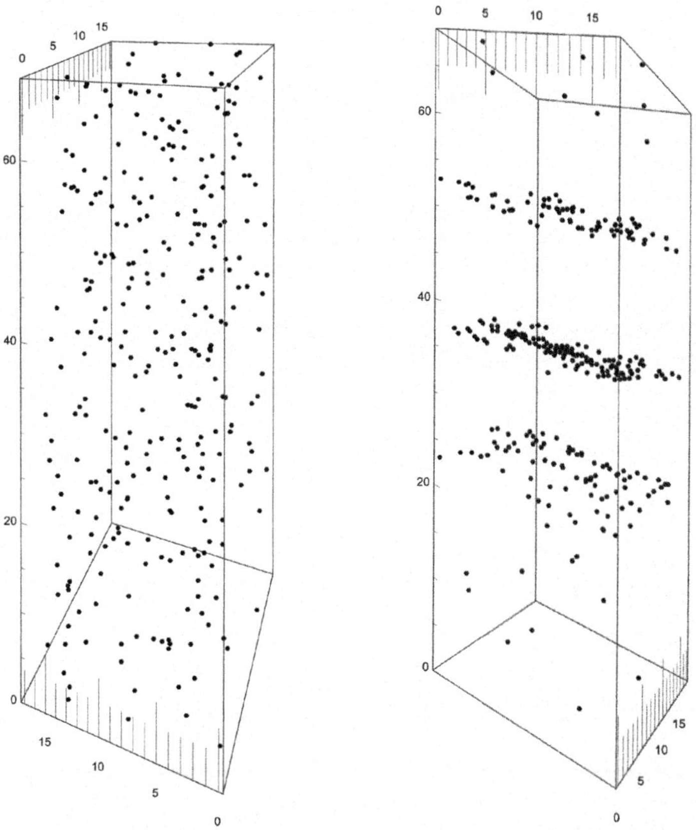

Fig. 5.13 Simulations for the $d = 3$ dimensional reduction to $\mathbb{S}^1 \times \mathbb{S}^1 \times I$ with $n = 300$ elements for $\beta < \beta_c$ and $\beta > \beta_c$. The phase on the left is random and continuumlike and that on the right is layered and non-continuumlike

all other elements (this is the **originary** condition we will learn about in Sect. 5.3.) The Hartle-Hawking wave function is then a functional over the future-most and hence inextendible antichains $\mathcal{A}_f$ of the causal sets in the ensemble, with some given (intrinsic) geometric properties. The simplest property to impose is of fixed cardinality,[1] i.e., $|\mathcal{A}_f| = \mathcal{N}_f$. We have seen in Chap. 3 that for continuumlike causal sets this is proportional to the spatial volume of a corresponding spatial hypersurface Σ. The causal set Hartle-Hawking(HH) wave function for the set $\Omega_{HH}(\mathcal{N}_f, n)$ of n-element causal sets with $(\mathcal{N}_i = 1, \mathcal{N}_f)$ is defined as

[1] One could also use a thickened antichain to capture the boundary geometry as in Chap. 3. Given computational restrictions, as well as the fact that the markov chain can traverse non-continuumlike causal sets for which the identification is less meaningful, we will stick to this minimalist definition of spatial geometry.

$$\Psi_0^{(n)}(N_f, \beta) \equiv A \sum_{\Omega_{HH}(N_f,n)} e^{-\beta S(C)}. \tag{5.32}$$

For $\beta = 0$, $\Psi_0^{(n)}(N_f, 0)$ is therefore the (normalised) uniform distribution over $\Omega_{HH}(N_f, n)$.[2]

The number of causal sets in $\Omega_{HH}(N_f, n)$ depends on n_b, the number of bulk elements, where $1 + N_f + n_b = n$. Thus for the largest values of N_f, n_b is very small and the HH wavefunction can be evaluated analytically. This is convenient since, for the largest values of N_f, the MCMC simulations become less reliable, with thermalisation times becoming significantly larger.

We can perform the sum Eq. (5.32) analytically for the largest values of N_f. Since the element in $\mathcal{A}_i$ is to the past of all other elements it can be uniquely labelled as $e_0 = (0, 0)$. For $N_f = n - 1$ or $n_b = 0$ this is simple, since $\Omega_{HH}(n-1, n)$ consists of a single causal set with e_0 linked to all the elements in $\mathcal{A}_f$, i.e., $N_0 = n - 1$. This gives us the simple expression

$$\Psi_0(n-1, \beta) = Ae^{-\beta 2\epsilon n(1-2\epsilon)+4\epsilon^2}. \tag{5.33}$$

When $N_f = n - 2$, there is a single element in the bulk, $e_1 = (u_1, v_1)$ which is to the future of e_0 and to the past of at least one element in $\mathcal{A}_f$ as shown in Fig. 5.14. Every causal set in $\Omega_{HH}(n-2, n)$ can thus be characterised by the number of elements l to the future of e_1, where $l \in (1, \ldots, n-2)$. For each l, $N_0 = n - 1$, $N_2 = l$, while $N_i = 0$, $\forall i > 2$, and so the action depends only on l, and we can reduce the sum Eq. (5.32) to a sum over l. Each l term in this sum gets a contribution from μ_l causal sets, where μ_l is calculated as follows.

For any $e = (u, v) \in \mathrm{Fut}(e_1)$, $u > u_1$ and $v > v_1$ and hence, if there are l such elements, this means that u_1, v_1 are constrained to take values in the smaller set $[1, \ldots n - 1 - l]$. Moreover, if $e_s = (u_s, v_s)$, is one of the $N - 2 - l$ elements in $S(e_1) \equiv \mathcal{A}_f \backslash \mathrm{Fut}(e_1)$, and if $u_s < u_1$, then $v_s > v_1$, else if $u_s > u_1$ then $v_s < v_1$. The number of such configurations for a fixed $\mathrm{Fut}(e_1)$ is therefore $u_1 - 1 + v_1 - 1$ and should equal the number of elements in $S(e_1)$, i.e., $v_1 + u_1 = n - l$. Let the elements $e_f \equiv (u_f, v_f)$ in $\mathrm{Fut}(e_1)$ be labelled by ordering the u_f values $u_{i_1} < u_{i_2} < \ldots < u_{i_l}$. Since $\mathrm{Fut}(e_1) \subset \mathcal{A}_f$ is an antichain, $v_{i_1} > v_{i_2} > \ldots > v_{i_l}$. For a fixed u_1 therefore the labelling of all other elements is fixed and hence $\mu_l = (n - 1 - l)$. Thus

$$\Psi_0(n-2, \beta) = \frac{Ae^{-\beta(R-Q)}}{(1-e^{\beta Q})^2}\left(n - 2 - (n-1)e^{\beta Q} + e^{\beta Q(n-1)}\right) \tag{5.34}$$

where $Q = 4\epsilon^2(1 - 3\epsilon)$ and $R = 2\epsilon n(1 - 2\epsilon) + 4\epsilon^2$ is given as in (5.33).

Similar arguments can be used to calculate $\Psi_0(n-3, \beta)$ for $N_f = n - 3$ [184], but as n_b increases so does the number of distinct configurations in the bulk and the

[2] We have suppressed the $\hbar$ in the above expression, but the vigilant reader should keep track of it when needed!

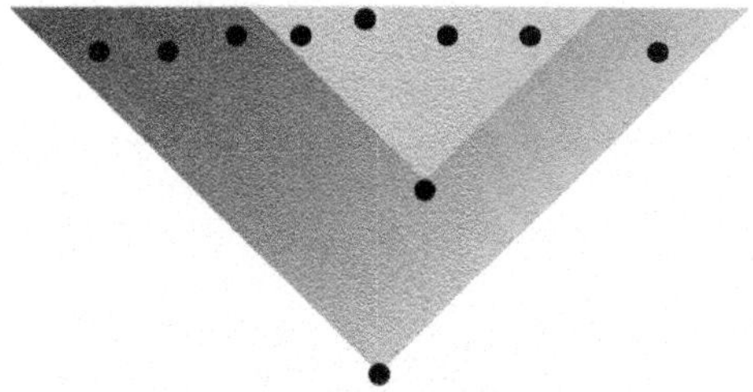

Fig. 5.14 An example of a causal set with $\mathcal{N}_f = n - 2$. There is a single bulk element which is to the future of $\mathcal{A}_i = \{e_0\}$, and must be to the past of at least one of the elements in $\mathcal{A}_f$

calculation requires an explicit enumeration of posets, which as we have seen is only known upto $n = 16$!

Instead, $\Psi_0(\mathcal{N}_f, \beta)$ can be calculated using the MCMC [184] methods described earlier. These calculations give us thermalised expectation values of the action $\langle S(\beta, \mathcal{N}_f)\rangle$ for all values of $\mathcal{N}_f$, and by varying β, we can also numerically find its dependence on β. This allows us to perform the numerical integration $\int_0^\beta d\beta' \langle S(\beta', \mathcal{N}_f)\rangle$ using which we reexpress Eq. (5.32) as

$$\Psi_0(\mathcal{N}_f, \beta) = A\mathcal{Z}_\beta(\mathcal{N}_f) = A\mathcal{Z}_0(\mathcal{N}_f)\exp\left(-\int_0^\beta d\beta' \langle S(\beta', \mathcal{N}_f)\rangle\right). \tag{5.35}$$

The normalised functional $\Psi_0(\mathcal{N}_f, \beta)$ depends on the factor $\mathcal{Z}_0(\mathcal{N}_f)$, which must therefore be explicitly calculated.

The unrestricted $\beta = 0$ partition function is

$$\mathcal{Z}_0 = \sum_{\mathcal{N}_f} \mathcal{Z}_0(\mathcal{N}_f), \tag{5.36}$$

which is dominated by 2d random orders. Upto an overall normalisation, it is a sum of frequencies of $\mathcal{N}_f$ in the ensemble of 2d random orders and hence can be numerically estimated for large enough $\mathcal{N}_f$. This can be supplemented with the analytic values for large $\mathcal{N}_f$ and hence can be used to normalise $\Psi_0(\mathcal{N}_f, \beta)$. This was the procedure adopted in [57], but since there were intermediate values of $\mathcal{N}_f$ which were amenable to neither calculation, the functional form was obtained via an interpolation.

MCMC simulations in [57] show both similarities as well as differences with the unrestricted path sum for the 2d orders. Fixing $\mathcal{N}_f$ we see the same behaviour as in the 2d orders. For small values of $\mathcal{N}_f$ there is a phase transition similar to that for 2d orders with a continuum phase for $\beta < \beta_c$ and a highly connected non-continuum "crystalline" phase for $\beta > \beta_c$. As $\mathcal{N}_f$ increases, however, the critical point β_c first begins to decrease, achieving a minimum value β_c^{min} after which it begins to increase again. This feature turns out to be essential for determining the final form of $\Psi_0(\mathcal{N}_f, \beta)$.

After obtaining $\langle S(\beta, \mathcal{N}_f)\rangle$ for all the values of $\mathcal{N}_f$ as well as as a large number of β values, the numerical integration was performed to get

$$\ln \frac{\mathcal{Z}_0(\mathcal{N}_f)}{\mathcal{Z}_\beta(\mathcal{N}_f)} = \int_0^\beta d\beta' \langle S(\beta', \mathcal{N}_f)\rangle. \tag{5.37}$$

Using the normalisation condition $\Sigma_{\mathcal{N}_f=1}^{N-1}|\Psi_0(\mathcal{N}_f)|^2 = 1$, gives us the form for the Hartle-Hawking functional $\Psi_0(\mathcal{N}_f)$.

Figure 5.15 shows the results for $n = 50$. It is striking that $\Psi_0(\mathcal{N}_f)$ forms well-defined peaks about specific discrete geometries and that the peak shifts discontinuously as one passes β_c. It is important to emphasise here that the well-formed peaks at all values of β is not a result of tweaking of parameters, or the assumption of a coherent state, or a symmetry reduction. This is genuine, fully non-perturbative bottom-up numerical calculation of the quantum gravity Hartle-Hawking wave function. Without any extra assumptions it is clearly peaked on two very special discrete geometries.

The first peak at $\beta < \beta_c$ corresponds to a distinctly continuum phase while the second peak at $\beta > \beta_c$ corresponds to a non-continuum phase (see Fig. 5.16) with a distinctive layered structure characteristic of the highly connected phase that we've already observed. Since the free parameter β has a ready interpretation as an inverse temperature, as in the Euclidean path integral, the dominance at large β of the non-continuumlike causets thus could be taken to mean that they represent the ground state of the theory. While they have no continuum counterpart, they nevertheless possess properties that are of particular significance to the early universe:

- The ratio of $\mathcal{N}_f$ to the height of the causal set at the second peak shows a rapid expansion from a single initial element to a large final antichain.
- Each element in the final antichain is causally connected to most of the elements in the previous layer, in stark contrast with the conformal causal structure of the standard FRW universe.

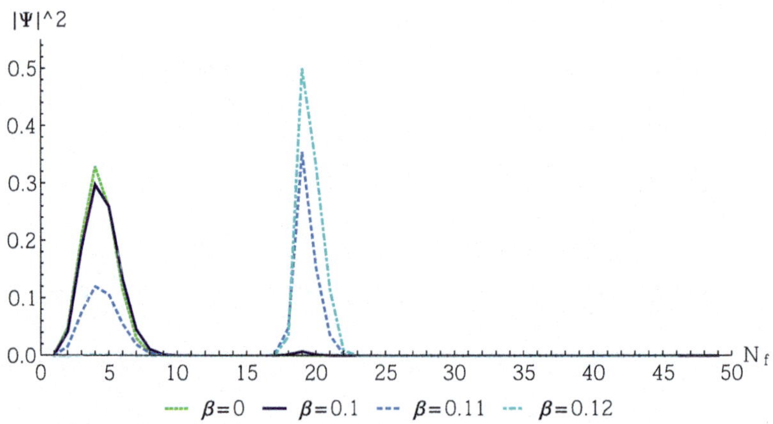

Fig. 5.15 $|\Psi_0(\mathcal{N}_f)|^2$ for $\epsilon = 0.5$. For $\beta < \beta_c$ there is a stable peak around $\mathcal{N}_f \sim 5$, which jumps into a stable peak at around $\mathcal{N}_f \sim 20$ for $\beta > \beta_c$. As β_c is approached from below, a small secondary peak develops at $\mathcal{N}_f \sim 20$ and similarly from above, the secondary peak develops at $\mathcal{N}_f \sim 5$. The dominant peak nevertheless jumps at the phase transition. This is also suggestive of a first order phase transition where both phases exist simultaneously at the phase transition, and is manifested here by the behaviour of the HH wave function

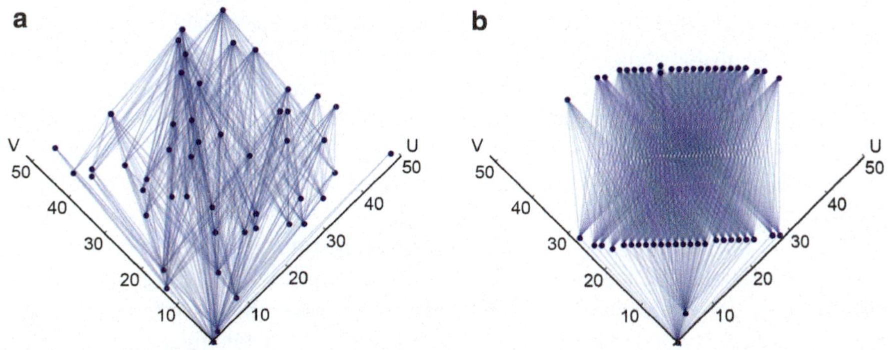

Fig. 5.16 Typical 2d orders at $\epsilon = 0.12$ associated with **a** the first and **b** the second peak, in lightcone coordinates. The lines indicate causal relations between elements

- The final antichain is homogeneous: the past of each element is almost the same for all the elements in the antichain.

Although this is just a $d = 2$ system, this is surprisingly close to the initial conditions for our own universe and suggests alternatives to continuum inflationary scenarios. While the restriction to 2d is clearly unphysical, as in other approaches one hopes to learn general lessons from this. An important and open problem in causal set theory is to generalise this calculation to higher dimensions. A technical issue is that the normalisation procedure is no longer quite so straightforward, but is perhaps possible to perform in principle.

5.3 The Sequential Growth Paradigm

Although the discrete path integral seems a reasonable way to construct the quantum dynamics of causal sets, it is continuum-inspired and not built from a more fundamental set of order-theoretic principles. In particular, while the discrete Einstein-Hilbert or Benincasa-Dowker-Glaser (BDG) action is defined order theoretically, the form—both the coefficients as well as the dimension—are obtained by comparison to the continuum. One option is of course to generate the terms in the action using a renormalisation procedure, as suggested in [186], but this again takes cues from continuum-inspired arguments.

The sequential growth paradigm (SGP) of Rideout and Sorkin is a stochastic growth model where the causal sets are "grown" element by element, starting from an initial single element. The classical sequential growth (CSG) models satisfy a set of physical principles, including causality, and are completely determined by a single coupling constant at each stage of growth [62,63] and have been well studied [62–65,196–201]. This has led to several interesting philosophical discussions on whether these models belong to the category of so-called growing block-universes or not [202–204]. A discussion on these issues is well outside the scope of this book and we refer the reader to the original sources.

The stochastic dynamics generates the set of past finite causal sets $\Omega_{\rm pf}$ and is naturally cast in the language of measure theory. It mimics many aspects of the discrete random walk, with label invariant "events" or measurable subsets generating the sigma algebra $\mathfrak{S}$ over the sample space $\Omega_{\rm pf}$. These measurable events take the place of covariant observables [65] or beables (in Bell's phraseology [205]) for CST. The dynamics is given by the measure triple $(\Omega_{\rm pf}, \mathfrak{S}, \mu)$ where μ is a probability measure in CSG, generated from the couplings.

The quantum version of this dynamics, the **quantum sequential growth (QSG)** models, is a quantum generalisation of this measure triple, with μ being replaced by a **quantum measure**, which can be thought of either as a **decoherence functional**, or a **vector measure**, valued in a histories Hilbert space. Such a dynamics is not trivial to construct and faces two main challenges. The first is to find a consistent generalisation of the physical principles used in CSG which incorporates entanglement suitably. The second is that the finitely generated quantum measure need not be defined over covariant events—for this it needs to satisfy special convergence conditions [70,71].

For certain choices of the couplings, there are hints of a continuumlike behaviour from CSG models [200] but it is fair to say that little is known about the continuum approximation in general. It is likely that emergent dynamics will require new renormalisation techniques (see [64]) and coarse graining, but for now the focus is to obtain conditions for a consistent quantum dynamics.

5.3.1 The Sequential Growth Framework

Constructing a dynamics framework for quantum gravity is tricky—how can we define an evolution when there is no covariantly defined time parameter? In the growth paradigm for causal sets we will describe how to define a natural "clock" for causal sets with which to describe dynamics. While the clock is inherently non-covariant, the framework nevertheless allows us to obtain the full set of covariant observables. Unlike other approaches which use clocks intrinsically, the sequential growth clock is considered to be unphysical, and hence can be "gauged away".

In the continuum, the past spacetime volume (say up to an initial hypersurface) is a possible choice of time parameter. As we have learnt, in the continuum approximation the cardinality of a causal set corresponds to the spacetime volume in the continuum. Thus, the "growth" or increase in the cardinality of a causal set is analogous to a ticking clock.

Starting from an n-element causal set we can run such a growth backward to arrive at the natural starting point of zero cardinality, the empty set, from which the growth process starts. There are of course multiple ways to grow a poset—to the future, to the past, or both [206,207]. In either case, the growth produces causal sets whose elements are labelled by the growth process. Following [207] we define a **natural labelling** or a **linear extension** of a causal set C as the bijection ℓ from C to either $\mathbb{N}$, $\mathbb{Z}^-$ or $\mathbb{Z}$ such that $\ell(i) \prec \ell(j) \to i < j$. Henceforth we will use the terminology "labelled causal" set to mean one that is naturally labelled.

Let us first consider the growth to the future. In this case, every new element must lie either to the future of an existing element or be unrelated to it, i.e., the future of the causal set changes, not its past. This satisfies the principle of **internal temporality**, where the sequence of the growth marks the ticking of the clock. Figure 5.17 shows the growth process upto $n = 3$, which generates a poset of naturally labelled causal sets, **Poscau** $\mathcal{P}$. The natural labelling of such posets is given by $\mathbb{N}$ [207] where at **stage** n, an $n + 1$ element causal set has been grown, with the new element e_{n+1} added at stage $n + 1$. Such a growth can only generate labelled causal sets that are **past finite** since the past of every element e_n can at most have $n - 1$ elements. For now, for notational consistency, we will denote the set of all countable past finite, labelled posets by $\widetilde{\Omega}_{\mathbb{N}}$.

The backward evolution is analogously defined [207]: the new element must be added to the past of an existing element or be unrelated to it. In this case, the posets are naturally labelled by $\mathbb{Z}^-$, and as one runs $n \to -\infty$ this generates the set of **future finite** countable causal sets $\widetilde{\Omega}_{\mathbb{Z}^-}$. This backward growth process also satisfies a generalised principle of internal temporality, with the sequence of the growth marking the ticking of a clock running backwards in time, by ever decreasing values of $\mathbb{Z}^-$.

The third option of both forward and backward growth is more complicated with elements being added to the future and/or the past [207]. In this case, the posets are natural labelled by $\mathbb{Z}$, i.e., with each element labelled by either a positive or a negative integer. These "alternating growth" models were constructed explicitly in [207] where the growth proceeds by the alternating sequence $0, -1, 1, -2, 2, -3, 3$, etc. Thus at odd stage n, the -$(n + 1)/2$th element is added, and at even stage n, the $n/2$th element is added. Again, a generalised internal temporality is satisfied since positive elements cannot be added to the past of an existing element and negative elements cannot be added to the future of existing elements. This leaves intact the causal set $\widetilde{C}_{n-1}$ that has been grown upto stage $n - 1$, since the new element cannot be inserted into an Alexandrov interval that has been formed in $\widetilde{C}_{n_1}$. The set of causal sets generated as $n \to \pm\infty$ is denoted by $\widetilde{\Omega}_{\mathbb{Z}}$: these are neither past or future finite,

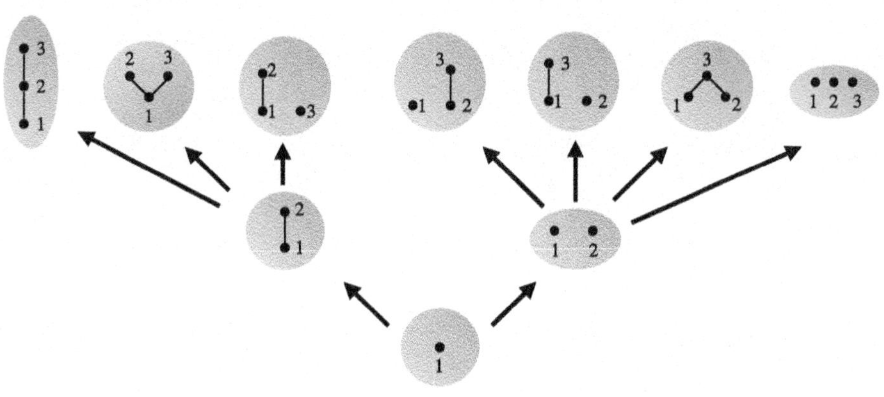

Fig. 5.17 The forward labelled sequential growth tree, labelled **Poscau** $\widetilde{\mathcal{P}}$ up to stage $n = 3$. Three of the 3-element causal sets are order isomorphic

but nevertheless satisfy one of the defining conditions of causal set, namely local finiteness.

Because of the labelling, the sets $\widetilde{\Omega}_{\mathbf{N},\mathbb{Z}^-,\mathbb{Z}}$ are all pair-wise mutually disjoint and hence the set of all labelled infinite element causal sets is given by their disjoint union: $\widetilde{\Omega} = \widetilde{\Omega}_{\mathbf{N}} \sqcup \widetilde{\Omega}_{\mathbb{Z}^-} \sqcup \widetilde{\Omega}_{\mathbb{Z}}$. We denote the unlabelled counterparts of $\widetilde{\Omega}_{\mathbf{N},\mathbb{Z}^-,\mathbb{Z}}$ by the equivalence class $\Omega_{\mathbf{N},\mathbb{Z}^-,\mathbb{Z}}$ where $\widetilde{C} \sim \widetilde{C}'$ under relabellings of C.

In each of these cases—future, past or alternating—the growth kinematics is captured by a tree of all possible labelled causal sets that are generated. Let us consider the future growth model first. Every n-element node in the tree is an n-element labelled causal set $\widetilde{C}$ and is generated from a unique labelled **parent** $\widetilde{C}_{n-1}$. Thus, the sub-causal set $\widetilde{C}_n|_{n-1}$ obtained from restricting to the first $n-1$ elements of $\widetilde{C}_n$ is order isomorphic to $\widetilde{C}_{n-1}$. This generates labelled Poscau $\widetilde{\mathcal{P}}_{\mathbf{N}}$ which is a tree of labelled causal sets. One can similarly define the labelled poscau for the past growth $\widetilde{\mathcal{P}}_{\mathbb{Z}^-}$ and the alternating growth model $\widetilde{\mathcal{P}}_{\mathbb{Z}}$.

In this book we will focus our attention on the future growth models which have been studied most extensively, starting with the Rideout-Sorkin models [62–65,196,199,201,208,209]. For these models, the sample space can be expressed as the set of all countable past finite causal sets $\Omega_{\mathbb{N}} = \Omega_{\mathrm{pf}}$. The past and future-past extended sequential growth models can be found in [207].

5.3.1.1 A Random Walk Refresher

Since much of the measure theoretic framing of the sequential growth models rests on the same mathematical structure as the random walk (RW) it is worthwhile to start with a review of the RW in the $d = 2$ spacetime lattice $\mathbf{L}_2$.

Consider the RW which starts at the spatio-temporal origin $(t = 0, x = 0)$. We can think of this point as a 1-element **history** or **path** $\gamma_0 = \{(0, 0)\}$ in $\mathbf{L}_2$. At $t = 1$, the walker can be either at $(0, 1)$ or $(0, -1)$. The corresponding path they take is now a 2-path, either $\gamma_1^+ \equiv \{(0, 0), (1, 1)\}$ or $\gamma_1^- \equiv \{(0, 0), (1, -1)\}$, which are "sequentially grown" from γ_0. The RW thus gives rise to a poset, or more specifically a tree of paths $\mathcal{P}_{\mathbf{RW}}$, where each element of $\mathcal{P}_{\mathbf{RW}}$ is an n-path in $\mathbf{L}_2$ (see Fig. 5.18.)

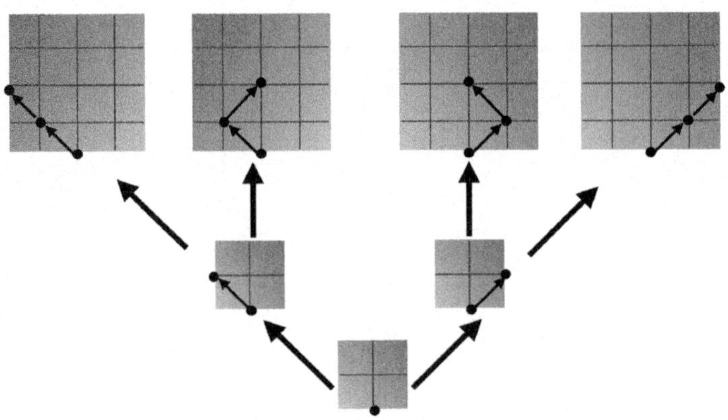

Fig. 5.18 Tree of paths $\mathcal{P}_{\mathbf{RW}}$ for the Random Walk up to stage $T = 2$

At every node γ_n of $\mathcal{P}_{\mathbf{RW}}$ there are only two ways to grow to an $n+1$-path; the random walker having traversed the path γ_n upto time $t_n = n$ can choose to take the step $(t_{n+1}, x_n + 1)$ or $(t_{n+1}, x_n - 1)$, where (t_n, x_n) is the future-most point on γ_n. The tree $\mathcal{P}_{\mathbf{RW}}$ has exactly 2^n number of n-element nodes since it branches exactly twice at every node. Note that the branching number in $\widetilde{\mathcal{P}}_{\mathbf{N}}$ is far more complex and depends not only on n, but also the details of the node itself. As we learnt from Sect. 5.2.1, the number of n-element nodes grows super-exponentially as $2^{n^2/4+o(n)}$, which makes the structure of $\widetilde{\mathcal{P}}_{\mathbf{N}}$ far more complex than that of $\mathcal{P}_{\mathbf{RW}}$.

The RW on $\mathbf{L}_2$ can thus be equivalently viewed as an RW on $\mathcal{P}_{\mathbf{RW}}$. If p denotes the probability for the step $(t_n, x_n) \rightarrow (t_{n+1}, x_n + 1)$, the transition probability from the parent node γ_n to its child node γ_{n+1} is either p or $(1-p)$, depending on whether the future most point on γ_{n+1} is $(t_{n+1}, x_n + 1)$ or $(t_n, x_n - 1)$. One can therefore "dress" $\mathcal{P}_{\mathbf{RW}}$ with these probabilities. The probability for a given node γ_n is then the product of transition probabilities on the unique path from γ_0 to γ_n in $\mathcal{P}_{\mathbf{RW}}$

$$\mathrm{Prob}(\gamma_n) = p^k(1-p)^{n-k} \tag{5.38}$$

where k is the number of times γ_n goes to the right, starting from $(0, 0)$. This gives us a probability measure on the set of all finite time paths γ_n which is also *finitely* additive, and extends to sets of finite time paths. For any set of $j \leq 2^n$ *distinct* n-element paths $\Gamma \equiv \{\gamma_n^1, \ldots, \gamma_n^j\}$

$$\mathrm{Prob}(\Gamma) = \sum_{i=1}^{j} \mathrm{Prob}(\gamma_n^i). \tag{5.39}$$

We may also want to calculate the measure on sets of varying sized paths. For example, the **event** that the random walker comes back to the origin by a finite time T is the set of all finite time paths for which the random walker returns to the origin for the first time at $n \leq T$. The first return could happen at any even time $n = 2, 4, \ldots, \leq T$. For a first return at time $n = 4$ for example, all subsequent steps of the walker are irrelevant, and hence the probability for this event is itself a sum over probabilities over all 5-element paths which have returned to the origin at time 4. This means the event of a first return at any even time n is itself a set of $n+1$-element paths.

Apart from such finite time questions, one may also be interested in infinite-time events. These are particularly important for causal sets since it is only via such events that covariant events can be defined. An example is the event that the random walker returns to the origin. Since the first return can happen at any time, one has to consider the set of paths with first return at any even $n < \infty$. The measure for such an event however, is a sum over a countable infinity of paths. Do such sums converge? Thanks to the Kolmogorov-Caratheodary-Hahn (KCH) theorem, the answer to this is in the affirmative, for probability measures.

We now come to the mathematical formulation of this important result in measure theory, since this structure is essential to our construction of the sequential growth dynamics for causal sets.

The **finite sample space** $\Omega_{\mathbf{RW}}(n) \equiv \{\gamma_n\}$ is the set of all finite n step paths and the full **sample space** of the RW is the space of all infinite time paths

$$\Omega_{\mathbf{RW}} \equiv \lim_{n\to\infty} \Omega_{\mathbf{RW}}(n). \tag{5.40}$$

with $\mathrm{Prob}(\Omega_{\mathbf{RW}}) = 1$. Since $\Omega_{\mathbf{RW}}$ contains a countable infinity of paths, unless the dynamics is deterministic ($p = 1$ or $p = 0$ and the particle moves at constant velocity either to the right or to the left, respectively), the probability for any single path must almost surely be zero. In order to make physical and mathematical sense of this, one therefore has to find the measureable sets in $\Omega_{\mathbf{RW}}$. The finite time paths come to the rescue: for every finite time path γ_n, $n < \infty$, define the **cylinder set**

$$\mathrm{cyl}(\gamma_n) \equiv \{\gamma \in \Omega_{\mathbf{RW}} |\ \gamma|_n = \gamma_n\} \subset \Omega_{\mathbf{RW}} \tag{5.41}$$

where $\gamma|_n$ denotes the restriction of the infinite time path γ to the first n-steps. Thus a cylinder set $\mathrm{cyl}(\gamma_n)$ is the set of all paths in $\Omega_{\mathbf{RW}}$ whose first n elements are the finite path $\gamma_n \in \Omega_{\mathbf{RW}}(n)$. On the tree $\mathcal{P}_{\mathbf{RW}}$, each node is therefore a cylinder set in $\Omega_{\mathbf{RW}}$. In Fig. 5.19 we show a cylinder set for the RW and for the sequential growth for causal sets.

Although $\mathrm{cyl}(\gamma_n)$ contains a countable infinity of paths in $\Omega_{\mathbf{RW}}$, it can be assigned a natural measure, namely, that of γ_n itself

$$\mu(\mathrm{cyl}(\gamma_n)) \equiv \mathrm{Prob}(\gamma_n) = p^k(1-p)^{n-k-1}, \tag{5.42}$$

where as before, k is the number of times γ_n goes to the right. On the $\mathcal{P}_{\mathbf{RW}}$, this gives the probability of the node γ_n.

The cylinder sets are measurable and hence can be used to generate the set of all finitely measurable sets. Let $\Gamma_{\mathbf{RW}}(n)$ denote the set of all cylinder sets of **size** n and

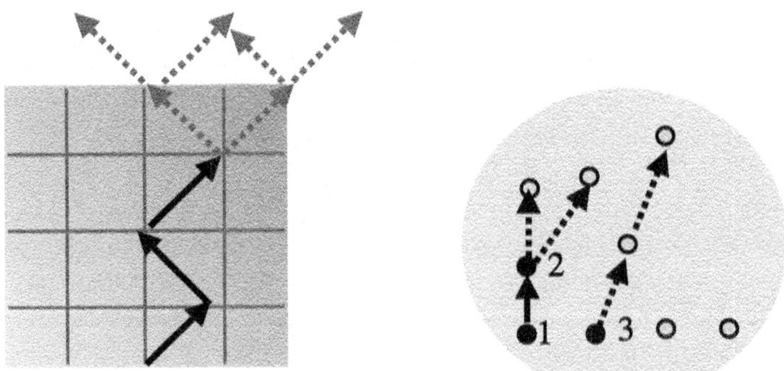

Fig. 5.19 On the left is the cylinder set $\mathrm{cyl}(\gamma_3)$ in the RW, which is the set of all possible paths that continue onward from γ_3 to infinity. On the right is a depiction of a cylinder set cyl( ) for sequential growth, which is the set of all countable past finite, naturally labelled causal sets with the first 3 elements being

$\Gamma_{\mathbf{RW}}$ the set of all cylinder sets. The **event algebra** $\mathfrak{A}$ of the RW is then generated, via finite set operations (unions, intersections and complementations) on $\Gamma_{\mathbf{RW}}(n)$ and includes the null set $\emptyset$ and $\Omega_{\mathbf{RW}}$ itself.

We list some important properties of RW cylinder sets which are also shared by the cylinder sets for sequential growth in causal set theory:

1. Distinct cylinder sets of size n do not intersect: for $\alpha_n, \beta_n \in \Gamma_{\mathbf{RW}}(n)$
$$\alpha_n \neq \beta_n \Rightarrow \alpha_n \cap \beta_n = \emptyset. \tag{5.43}$$
2. If two cylinder sets of differing sizes intersect non-trivially, then one must be strictly contained in the other. For $n < m$, $\alpha_n \in \Gamma_{\mathbf{RW}}(n)$, $\beta_m \in \Gamma_{\mathbf{RW}}(m)$
$$\alpha_n \cap \beta_m \neq \emptyset \Rightarrow \beta_m \subset \alpha_n. \tag{5.44}$$
We will refer to this as the **nesting property**.
3. For any γ_n,
$$\mathrm{cyl}(\gamma_n) = \mathrm{cyl}(\gamma_{n+1}^+) \cup \mathrm{cyl}(\gamma_{n+1}^-). \tag{5.45}$$
4. Any element of $\Gamma_{\mathbf{RW}}$ can be expressed as a **fine partition**, i.e., as a finite union of cylinder sets of the same size. For $\alpha \in \Gamma_{\mathbf{RW}}$, $\exists\, n < \infty$ and a subset $\Gamma_{\mathbf{RW}}(n, \alpha) \subset \Gamma_{\mathbf{RW}}(n)$ such that
$$\alpha = \bigcup_{\alpha_n \in \Gamma_{\mathbf{RW}}(n,\alpha)} \alpha_n. \tag{5.46}$$
There is moreover a unique coarsest fine partition, i.e. a smallest $n(\alpha)$ for which a fine partition of α exists.

Exercise 5.7 Prove Eqs. (5.43)–(5.46). Use the tree $\mathcal{P}_{\mathbf{RW}}$ as a guide.

Using Eq. (5.43) we note that the finite sum in Eq. (5.45) is over mutually disjoint sets and hence we can define, for every $\alpha \in \mathfrak{A}$

$$\mu(\alpha) \equiv \sum_{\alpha_n \in \Gamma_{\mathbf{RW}}(n,\alpha)} \mu(\alpha_n). \tag{5.47}$$

Thus we see that the measure on finite time paths is not only a measure of subsets of infinite time paths, but also extends to the entire finite time algebra $\mathfrak{A}$.

So far, the set operations we have defined are finite. In particular, the infinite time question of return to the origin requires one to perform countable set operations: if $R(n)$ denotes the set of paths that return to the origin for the first time at time n, then the set of return paths is

$$R \equiv \bigcup_{n=1}^{\infty} R(n). \tag{5.48}$$

That R does not belong to $\mathfrak{A}$ seems a mere technicality; since the $R(n)$ are disjoint, is it not straightforward to apply the additivity of probabilities to obtain a (hopefully) convergent sum $\mathrm{Prob}(R) = \sum_{n=1}^{\infty} \mathrm{Prob}(R(n))$? Alternatively, defining the set of paths that return for the first time to the origin by time $n-1$

$$R(\leq n) \equiv \bigcup_{m=1}^{n} R(m), \Rightarrow R = \lim_{n\to\infty} R(\leq n), \tag{5.49}$$

one might take $\mathrm{Prob}(R) = \lim_{n\to\infty} \mathrm{Prob}(R(\leq n))$. Do these two different routes give the same $\mathrm{Prob}(R)$?

The reason physicists don't have to worry about this question is because of the Kolmogorov-Caratheodary-Hahn (KCH) extension theorem referred to above, which guarantees that there is unconditional convergence of such sums and limits. The KCH theorem thus guarantees that the probability measure on $\mathfrak{A}$ "extends" to infinite time events. The set of all possible infinite time events that are generated by $\mathfrak{A}$ is the **sigma algebra** $\mathfrak{S}$ which is the smallest event algebra which contains $\mathfrak{A}$ and is closed under countable set operations on elements of $\mathfrak{A}$. The technical question that the KCH theorem answers is that the probability measure defined over $\mathfrak{A}$ extends to a *unique* probability measure on $\mathfrak{S}$ [210].

We have expressed the RW as a probability measure space, which is the measure triple $(\Omega_{\mathbf{RW}}, \mathfrak{A}, \mu)$, where

1. $\Omega_{\mathbf{RW}}$ is the sample space of infinite time paths,
2. a $\mathfrak{A}$ is the event algebra over $\Omega_{\mathbf{RW}}$ generated from finite set operations on cylinder sets
3. $\mu : \mathfrak{A} \to [0, 1]$ is a probability measure given by Eqs. 5.42 and 5.47.

By the KCH theorem, μ extends to a measure $\bar{\mu}$ on the sigma algebra $\mathfrak{S}$ generated by $\mathfrak{A}$ and hence $(\Omega_{\mathbf{RW}}, \mathfrak{A}, \bar{\mu})$ is also a probability measure space. Note that in the quantum case, where probabilities are replaced by analogues of amplitudes, the question of extension is not guaranteed. We will discuss this in detail in Sect. 5.3.3.

It would be useful to keep the RW in mind as we build up a sequential growth dynamics for causal sets in the subsequent sections.

5.3.1.2 The Sequential Growth Event Algebra and Covariant Beables

The sequential growth event algebra is constructed from $\widetilde{\mathcal{P}} = \widetilde{\mathcal{P}}_{\mathbb{N}}$ in a similar manner to the RW. Every node of $\widetilde{\mathcal{P}}$ is a finite labelled causal set $\widetilde{C}_n$, which can be thought of as a "proto" spacetime history of finite volume. $\widetilde{\mathcal{P}}$, while still a tree, has a branching number at every node that is determined not only by n, but also by the node $\widetilde{C}_n$ itself. At every stage n, the number of nodes grows, with the large n behaviour being super-exponential $2^{n^2/4+o(n)}$ [20]. Because of the complexity of the branching, we denote by $\Xi(\widetilde{C}^i_n) \equiv \{\widetilde{C}^{j(i)}_{n+1}\}$ the set of **children** of the **parent** $\widetilde{C}^i_n$ in $\widetilde{\mathcal{P}}$, i.e., the set of $n+1$ element causal sets emanating from the $\widetilde{C}^i_n$ node in $\widetilde{\mathcal{P}}$. The functional notation $j(i)$ is used as a short form to denote that j is valued in an index set $\mathfrak{I}(i, n) \subset \mathfrak{I}(n)$ of cardinality $|\Xi(\widetilde{C}^i_n)|$, which depends on $\widetilde{C}^i_n$. The simplest illustration of the varying node connectivity is at stage $n = 2$: the antichain $\widetilde{\mathrm{a}}_2$ has four children, while the chain $\widetilde{\mathrm{c}}_2$ has three children.

Exercise 5.8 Show that the number of labelled children of an n-element labelled antichain is 2^n, while that of an n-element labelled chain is $n + 1$.

As in the RW, the event algebra $\mathfrak{A}$ is generated by the cylinder sets,

$$\mathrm{cyl}(\widetilde{C}^i_n) \equiv \{\widetilde{C} \in \widetilde{\Omega}_{\mathrm{pf}} \mid \widetilde{C}|_n = \widetilde{C}^i_n\}, \ \mathrm{cyl}(\widetilde{C}^i_n) \subset \widetilde{\Omega}_{\mathrm{pf}} \tag{5.50}$$

associated with each node $\widetilde{C}^i_n$ in $\widetilde{\mathcal{P}}$, where $\widetilde{C}|_n$ denotes the first n elements of a labelled causal set $C \in \widetilde{\Omega}_{\mathrm{pf}}$. Even though the cardinality of $\mathrm{cyl}(\widetilde{C}^i_n)$ is not finite, we will refer to it as an **n-element cylinder set**.

Let $\mathcal{Z}_n$ denote the collection of n-element cylinder sets and $\mathcal{Z}$ the collection of all cylinder sets. As in the RW the event algebra $\mathfrak{A}$ is generated by taking finite unions, intersections and complements of the cylinder sets in $\mathcal{Z}$.

Again, the tree structure of $\widetilde{\mathcal{P}}$ ensures that the cylinder sets satisfy all four properties ((5.43)–(5.46)) of cylinder sets in the RW:

1. For a fixed n and any $\widetilde{C}^i_n \neq \widetilde{C}^j_n$

$$\mathrm{cyl}(\widetilde{C}^i_n) \cap \mathrm{cyl}(\widetilde{C}^j_n) = \emptyset \tag{5.51}$$

2. For $m < n$ and any $\widetilde{C}^i_m \in \widetilde{\Omega}_m$, $\widetilde{C}^j_n \in \widetilde{\Omega}_n$

$$\mathrm{cyl}(\widetilde{C}^i_m) \cap \mathrm{cyl}(\widetilde{C}^j_n) \neq \emptyset \Rightarrow \mathrm{cyl}(\widetilde{C}^j_n) \subset \mathrm{cyl}(\widetilde{C}^i_m). \tag{5.52}$$

3. For any $\widetilde{C}^i_n \in \widetilde{\Omega}_n$,

$$\mathrm{cyl}(\widetilde{C}^i_n) = \bigsqcup_{j(i)} \mathrm{cyl}(\widetilde{C}^{j(i)}_{n+1}). \tag{5.53}$$

4. For any $\alpha \in \mathfrak{A}$, there exists an integer $n_\alpha < \infty$ and a subset $S_\alpha \subset \{1, \ldots, \mathfrak{N}_{n_\alpha}\}$ such that

$$\alpha = \bigsqcup_{k \in S_\alpha} \mathrm{cyl}(\widetilde{C}^k_{n_\alpha}). \tag{5.54}$$

The **fine partition** of α is the set S_α with the smallest such n_α.

Figure 5.20 shows examples of cylinder sets as subsets of the sample space $\widetilde{\Omega}_{\mathrm{pf}}$. For $n = 2$, there are just two cylinder sets cyl() and cyl(). For $n = 3$, the cylinder sets cyl(), cyl(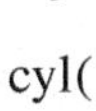), cyl() are the strictly non-overlapping subsets of cyl()

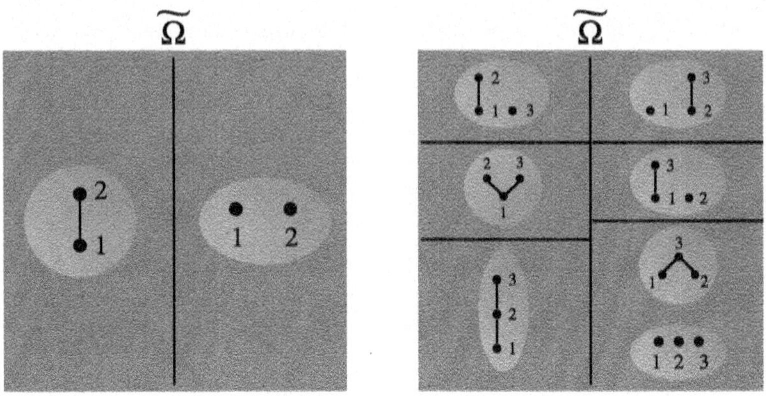

Fig. 5.20 **Cylinder sets** are subsets of the **sample space** $\widetilde{\Omega}_{\mathrm{pf}}$ of labelled past finite causal sets. The two $n = 2$ disjoint cylinder sets are shown on the left, and on the right *their* respective $n = 3$ disjoint cylinder subsets

where

$$\mathrm{cyl}({}^{2}_{1}) = \mathrm{cyl}({}^{3}_{2\,1}) \bigsqcup \mathrm{cyl}({}^{2\;3}_{\;1}) \bigsqcup \mathrm{cyl}({}^{2}_{1}\;{}_{3}), \tag{5.55}$$

while $\mathrm{cyl}({}^{3}_{1}\;{}_{2})$, $\mathrm{cyl}({}^{3}_{2}\;{}_{1})$, $\mathrm{cyl}({}_{1}{}^{3}{}_{2})$ and $\mathrm{cyl}(1\;2\;3)$ are the strictly non-overlapping subsets of $\mathrm{cyl}(1\;2)$ where

$$\mathrm{cyl}(1\;2) = \mathrm{cyl}({}^{3}_{1}\;{}_{2}) \bigsqcup \mathrm{cyl}({}^{3}_{2}\;{}_{1}) \bigsqcup \mathrm{cyl}({}_{1}{}^{3}{}_{2}) \bigsqcup \mathrm{cyl}(1\;2\;3). \tag{5.56}$$

The events in $\mathfrak{A}$ are labelled subsets of $\widetilde{\Omega}_{\mathrm{pf}}$ and a choice of dynamics (either classical or quantum) is a choice of measure on $\mathfrak{A}$. Unlike the RW, however, the labelled finite time events are not physical, i.e., they do not generate covariant subsets of $\widetilde{\Omega}_{\mathrm{pf}}$. A set $\alpha \subseteq \widetilde{\Omega}_{\mathrm{pf}}$ is said to be a **covariant subset** of $\widetilde{\Omega}_{\mathrm{pf}}$ if for every labelled $\widetilde{C} \in \alpha$, all its relabellings also belong to α, i.e.,

$$\alpha \equiv \{\widetilde{C} | \widetilde{C}' \sim \widetilde{C} \Longrightarrow \widetilde{C}' \in \alpha\}. \tag{5.57}$$

where $\widetilde{C} \sim \widetilde{C}'$ denotes an order isomorphism. A covariant subset α which also belongs to the event algebra is a **covariant event**. Thus a **covariant observable** or beable in sequential growth is a measurable covariant event.

We now show that no event in $\mathfrak{A}$ is covariant unless $\alpha = \widetilde{\Omega}_{\mathrm{pf}}$. For any $\alpha \in \mathfrak{A}, \alpha \neq \widetilde{\Omega}_{\mathrm{pf}}$, let $\mathcal{N}_\alpha$ be its fine partition Eq. (5.54). We can therefore write $\alpha = \bigsqcup_{k \in S_\alpha} \mathrm{cyl}(\widetilde{C}^k_{n_\alpha})$ where $S_\alpha \equiv |\mathcal{N}_\alpha|$. Let $\widetilde{C}^s_{n_\alpha}$ be a node (not necessarily unique) in $\mathcal{N}_\alpha$ with the largest number of minimal elements m_α. (i) Assume $n_\alpha > m_\alpha$, i.e., the n_α-antichain a_{n_α} does not belong to $\mathcal{N}_\alpha$. Let $\widetilde{C}^{g(s)}_{n_\alpha+1}$ denote the **gregarious child** of $\widetilde{C}^s_{n_\alpha}$, i.e., one in which the new element $e_{n_\alpha+1}$ is unrelated to all the elements in $\widetilde{C}^s_{n_\alpha}$ (the opposite is a **timid child**, where the new element $e_{n_\alpha+1}$ is to the future of all existing elements.) Then $\mathrm{cyl}(\widetilde{C}^{g(s)}_{n_\alpha+1}) \subset \mathrm{cyl}(\widetilde{C}^s_{n_\alpha})$ by the nesting property. Any $\widetilde{C}^l_{n_\alpha+1} \sim \widetilde{C}^{g(s)}_{n_\alpha+1}$ such that $\widetilde{C}^l_{n_\alpha+1} \neq \widetilde{C}^{g(s)}_{n_\alpha+1}$ cannot be a child of $\widetilde{C}^s_{n_\alpha}$ and hence $\mathrm{cyl}(\widetilde{C}^l_{n_\alpha+1}) \not\subset \mathrm{cyl}(\widetilde{C}^s_{n_\alpha})$. Moreover, since the number of minimal elements of $\widetilde{C}^l_{n_\alpha+1}$ is $m_\alpha + 1$, it cannot belong to the fine partition $\mathcal{N}_\alpha$ of α. Hence $\mathrm{cyl}(\widetilde{C}^l_{n_\alpha+1}) \cap \alpha = \emptyset$, or equivalently, for every $\widetilde{C} \in \mathrm{cyl}(\widetilde{C}^{g(s)}_{n_\alpha+1})$, there exists an order-isomorphic $\widetilde{C}' \in \mathrm{cyl}(\widetilde{C}^l_{n_\alpha+1})$ which does not belong to α. Hence α is not covariant. (ii) If $m_\alpha = n_\alpha$, then the antichain $\mathrm{a}_{n_\alpha} \in \mathcal{N}_\alpha$. Denote by $\mathcal{N}^c_\alpha$ the complement of $\mathcal{N}_\alpha$ in the set of n_α nodes.[3] If $m^c_\alpha < n_\alpha$ denotes the largest number of minimal elements for any node in $\mathcal{N}^c_\alpha$, then we can use (i) to show that $\alpha^c \in \mathfrak{A}$ is not a covariant event. Hence α is not a covariant event.

How should we build covariant events out of $\widetilde{\mathcal{P}}$? A hint was already given in the discussion in Sect. 5.3.1.1 on the infinite time events. Every $\alpha \in \mathfrak{A}$ admits a fine partion with a finite n_α and it is this latter property that makes the event non-covariant. If, for example, we were to take countable rather than finite unions of cylinder sets, such a fine partition, with a finite n_α will not generically exist.

An example of a covariant subset is the **originary event** $\alpha_{\mathrm{orig}} \subset \widetilde{\Omega}_{\mathrm{pf}}$ which is the set of causal sets for which there is a single element to the past of all the other elements in the causal set. This is a physically interesting subset of causal sets, roughly analogous to a big bang, as shown in Fig. 5.21. It is relatively easy to see that α_{orig} is invariant under natural relabellings.

Exercise 5.9 Consider the following two subsets in $\widetilde{\Omega}_{\mathrm{pf}}$: the set of causal sets for which there are at least $k > 0$ elements (a) with labelled past  and (b) with

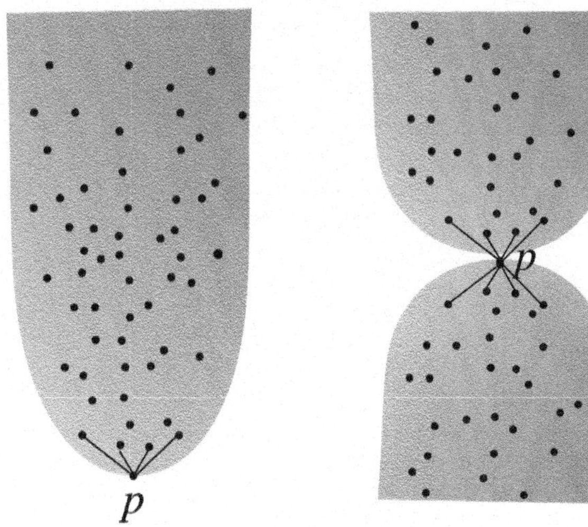

Fig. 5.21 An **originary** causal set in Ω_{pf} is depicted on the left where there is a single element p to the past of all other elements. The figure on the right depicts a **bounce** in a causal set, i.e., the existence of an element p which divides the causal set into its past and its future

[3] $\mathcal{N}^c_\alpha$ mus be non-empty, since otherwise $\alpha = \widetilde{\Omega}_{\mathrm{pf}}$.

unlabelled past . Are these covariant subsets? Explain.

α_{orig} can moreover be obtained from countable set operations on cylinder sets as follows. We begin with its complement in $\widetilde{\Omega}_{\text{pf}}$, namely the set of all causal sets that *do not* have a single element to the past of all elements. For any such causal set, somewhere in the growth process, there is a finite stage n at which the gregarious transition has been made, i.e., a gregarious child has been born. Since gregarious transitions can be made at *any* node $\widetilde{C}^i_n$, $\text{cyl}(\widetilde{C}^i_n)$ cannot itself be originary, even if $\widetilde{C}^i_n$ itself is. Any $\widetilde{C} \in \alpha^c_{\text{orig}}$ must therefore contain a 2-element subset a_2 of minimal elements. Just as in the return event, we need to consider the union over all cylinder sets $\text{cyl}(\widetilde{C}^i_n)$ where $n-1$ is the first stage at which a gregarious transition is made to $\text{cyl}(\widetilde{C}^i_n)$. Let I_n be the index set for the set of n-element nodes for which the nth element is the *only* gregarious element, i.e., the gregarious transition was taken for the first time at stage $n-1$. Using countable unions and complements, we see that the originary event can be expressed as

$$\alpha_{\text{orig}} = \left(\bigsqcup_{n>0} \bigsqcup_{i \in I_n} \text{cyl}(\widetilde{C}^i_n) \right)^c . \tag{5.58}$$

It should be obvious to the reader that α^c_{orig} bears more than a fleeting resemblance to the return event in the discrete random walk, which again uses countable unions of finite time events.

This suggests that covariant events can be constructed from cylinder sets only by performing countable set operations on the elements of $\mathfrak{A}$. The smallest algebra generated from all countable set operations from $\mathfrak{A}$ is the associated sigma-algebra $\widetilde{\mathfrak{S}}$ which is closed under all countable set operations on cylinder sets. Obviously not every event in $\widetilde{\mathfrak{S}}$ is covariant, so one must do further work to construct such events [65]. The set of covariant events themselves form a sub-sigma-algebra $\mathfrak{S}$ of $\widetilde{\mathfrak{S}}$ [65]. Despite being covariant $\mathfrak{S}$ is still a sigma algebra over the labelled sample space $\widetilde{\Omega}_{\text{pf}}$. In order to define covariant observables over the set of unlabelled causal sets Ω_{pf}, one must therefore consider the quotient algebra: $\mathfrak{S}/\sim$, where as before $\sim$ denotes the equivalence under relabellings.

A particularly important class of covariant events are the **stem events**, and the sigma algebra $\mathfrak{S}_{\text{stem}}$ they generate. A labelled **stem** $\tilde{s}$ in a labelled causal set $\widetilde{C} \in \widetilde{\Omega}_{\text{pf}}$ is a finite sub-causal set of $\widetilde{C}$, such that it contains its own past, in C, i.e., $\text{Past}(s) = s$. $\widetilde{C}$ is said to have an **unlabelled stem** s if there exists a natural labelling $\tilde{s}$ of s such that $\tilde{s}$ is a labelled stem in $\widetilde{C}$. An unlabelled $C \in \Omega_{\text{pf}}$ is said to have a stem s if s is a subset of C which contains its own past. These stem sets are therefore similar to certain past sets in continuum spacetimes, like the terminal indecomposable pasts (TIPs) of [211]. We define a **stem event** $\text{stem}(s)$ in $\widetilde{\mathfrak{S}}$ as the set of all labelled causal sets which contain the unlabelled stem s as a sub-causal set.

Exercise 5.10 Show that the stem event $\text{stem}(\widetilde{C}_n)$ is covariant.

An important example of a stem event is the complement of the originary event we have just described above, i.e., $\alpha^{c}_{\text{orig}} = \text{stem}(\text{a}_2)$, the set of causal sets that have the 2-element antichain a_2 as a stem.

Exercise 5.11 A **post** $p \in C \in \Omega_{\text{pf}}$ is an element such that $C = p \sqcup \text{Past}(p) \sqcup \text{Fut}(p)$. A **post event** α_{post} is the set of causal sets in Ω_{pf} which have at least one post. Show that α_{post} is not only covariant, but can be expressed in terms of countable set operations on stem events.

Posts are of particular interest because they correspond to points about which the universe "bounces" (see Fig. 5.21.) As the above exercise suggests, stem events can also be used to construct other covariant observables in $\mathfrak{A}$. The sigma-algebra $\mathfrak{S}_{\text{stem}}$ generated from the stem events is covariant and a sub-sigma algebra of $\mathfrak{S}$. For the class of Rideout-Sorkin dynamics we will discuss below in Sect. 5.3.2, those "rogue" events in $\mathfrak{S}$ which do not belong to $\mathfrak{S}_{\text{stem}}$ can be shown to be of zero measure and hence do not occur [65]. Thus, at least for these models, $\mathfrak{S}_{\text{stem}}$ captures all the physically realisable observables.

It is therefore of interest to ask whether stem events can be used to build a fully covariant growth model instead of labelled Poscau $\widetilde{\mathcal{P}}$. To begin with, one can construct the unlabelled Poscau $\mathcal{P}$, which is no longer a tree, but a more complicated poset. Here multiple routes to a single unlabelled causal set are possible. However, this does not by itself guarantee covariance, since the nodes continue to be labelled by stages.

In [199,201] a covariant tree, **Covtree** was defined using purely stem sets. Here, the nodes are specific collections of n-element stems that "certify" a $C \in \Omega_{\text{pf}}$ in the sense that these are the only n-element stems in C. Every n-element causal set therefore forms a node (they certify themselves!), but more importantly, so do certain subsets of n-element causal sets (but not all.) For $n = 2$ there are two 2-element causal sets: the 2-chain c_2 and the 2-element antichain a_2. One has therefore $\binom{2}{1} + \binom{2}{2} = 3$ combinations, i.e. three possible nodes: $\{\text{c}_2\}$, $\{\text{a}_2\}$ and $\{\text{c}_2, \text{a}_2\}$. While the first two nodes are "self certified", the latter is certified by the causal set . However, the complexity of the rule is already evident at stage $n = 3$, where there are five 3-element causal sets, which means that there are $\sum_{i=1}^{5} \binom{5}{i} = 2^5 - 1 = 31$ possible nodes, but only 22 of these can be certified [199,212].

Exercise 5.12 Show that the pair of stems $\{\text{c}_3, \text{a}_3\}$ do not form a node in Covtree.

Any dynamics on Covtree must be covariant, but it is a fairly challenging task to formulate a set of principles to determine this dynamics. In what follows we will therefore stay with Poscau $\widetilde{\mathcal{P}}$, and construct both classical and quantum dynamics on it.

5.3.2 The Rideout-Sorkin Classical Sequential Growth (CSG) Models

We begin with outlining the general (classical) measure theoretic framework in which the sequential growth paradigm naturally sits.

As in the RW, the dynamics is described by the measure triple $(\Omega, \mathfrak{A}, \mu)$ obtained from Poscau $\widetilde{\mathcal{P}}$. Here Ω is the sample space of past finite countable causal sets,[4] $\mathfrak{A}$ is the event algebra obtained from the cylinder sets and, in general, μ is any probability measure on $\mathfrak{A}$. By the KCH theorem, μ extends to a probability measure $\bar{\mu}$ on the labelled sigma algebra $\widetilde{\mathfrak{S}}$ generated by $\mathfrak{A}$, and also therefore to the covariant subalgebras $\mathfrak{S}$ and $\mathfrak{S}_{\text{stem}}$.

We wish to construct μ from a stochastic process on $\widetilde{\mathcal{P}}$ which respects the causal growth process. This means we must find a set of rules for defining transition probabilities on $\widetilde{\mathcal{P}}$, from a parent $\widetilde{C}_n^i$ to its child $\widetilde{C}_{n+1}^{j(i)}$, as one did for the RW. This results in a measure over finite time causal sets which are the nodes of $\widetilde{\mathcal{P}}$.

The Rideout-Sorkin models are described by the following set of transition rules, each of which is associated with a generalised physical principle.

1. **Markovian:** Let $\Xi(\widetilde{C}_n^i)$ denote the set of all children of the node $\widetilde{C}_n^i$ in $\widetilde{\mathcal{P}}$. Then the transition probabilities satisfy the sum rule

$$\sum_{j(i)} \mathbb{P}(\widetilde{C}_n^i \to \widetilde{C}_n^{j(i)}) = 1, \tag{5.59}$$

where $j(i)$ is valued in an index set $\mathfrak{J}(i,n)$ of cardinality $|\Xi(\widetilde{C}_n^i)|$. This means that, as in the RW, the measure on $\widetilde{C}_n^i$ is given by the product of transition probabilities on the unique path $\gamma(\widetilde{C}_n^i) \equiv \{\widetilde{C}_1^{k_1}, \widetilde{C}_2^{k_2}, \widetilde{C}_3^{k_3} \ldots \widetilde{C}_n^{k_n=i}\}$ in $\widetilde{\mathcal{P}}$ from the single element causal set $\widetilde{C}_1^{k_1} = \widetilde{C}_1$ to $\widetilde{C}_n^i$:

$$\mu(\widetilde{C}_n^i) \equiv \prod_{m=1}^{n-1} \mathbb{P}(\widetilde{C}_m^{k_m} \to \widetilde{C}_{m+1}^{k_{m+1}}) \tag{5.60}$$

2. **Covariance:** Let $\widetilde{C}_n^j$ be a natural relabelling of $\widetilde{C}_n^i$, i.e., $\widetilde{C}_n^j \sim \widetilde{C}_n^i$. The measure is invariant under the relabelling:

$$\mu(C_n^i) = \mu(\widetilde{C}_n^j). \tag{5.61}$$

This is equivalently a statement of path independence in the unlabelled Poscau $\mathcal{P}$. An example of this is shown in Fig. 5.17 where at stage $n = 3$ there are three order isomorphic nodes in $\mathcal{P}$, which then must be assigned the same measure.

[4] We drop the subscript on Ω_{pf} to simplify notation.

3. **Bell causality:** The measure satisfies **spectator independence**. Define the **spectator set** in a transition $\widetilde{C}_n^i \to \widetilde{C}_{n+1}^{j(i)}$ as the set of elements of $\widetilde{C}_n^i$ which are unrelated to the new element e_{n+1} in $\widetilde{C}_{n+1}^{j(i)}$, and the **precursor set** as the elements to the past of e_{n+1} in $\widetilde{C}_n^i$. For any non-timid transition, if we removed some or all of elements in the spectator sets from $\widetilde{C}_n^i$, we'd be left with a closely related transition $\widetilde{C}_m^{i'} \to \widetilde{C}_{m+1}^{j'(i')}$, where $n-m$ is the cardinality of the set of spectators that have been removed. These transition probabilities cannot be the same since this would violate the Markov condition, but we can incorporate spectator indepdendence by requiring that the relative probabilities of two transitions from $\widetilde{C}_n^i$ be independent of their mutually common spectator sets. Consider two non-timid transitions $\widetilde{C}_n^i \to \widetilde{C}_{n+1}^{j_1(i)}$ and $\widetilde{C}_n^i \to \widetilde{C}_{n+1}^{j_2(i)}$ with spectator sets S_1, S_2 respectively, such that $S_1 \cap S_2 \neq \emptyset$. Remove from $\widetilde{C}_n^i$ some non-zero subset $S_{12} \subset S_1 \cap S_2$ to get an m element causal set, $\widetilde{C}_m^k$. Since all one has done is remove a set of common spectators, one obtains the related transitions $\widetilde{C}_m^k \to \widetilde{C}_m^{l_1(k)}$, $\widetilde{C}_m^k \to \widetilde{C}_m^{l_2(k)}$, where $\widetilde{C}_m^{l_1(k)}$ and $\widetilde{C}_m^{l_2(k)}$ are obtained, respectively from $\widetilde{C}_m^{l_1(k)}$ and $\widetilde{C}_m^{l_2(k)}$ by removing S_{12}. Spectator independence or Bell Causality is the requirement

$$\frac{\mathbb{P}(\widetilde{C}_n^i \to \widetilde{C}_n^{j_1(i)})}{\mathbb{P}(\widetilde{C}_n^i \to \widetilde{C}_n^{j_2(i)})} = \frac{\mathbb{P}(\widetilde{C}_m^k \to \widetilde{C}_m^{l_1(k)})}{\mathbb{P}(\widetilde{C}_m^k \to \widetilde{C}_m^{l_2(k)})} \tag{5.62}$$

This condition can be reformulated as a product rule, which holds even when some of the transition probabilities are set to zero [196,209].

We note that while covariance ensures that the measure for a given $\widetilde{C}_n$ (and its associated cylinder set) is the same for all its order-isomorphic avatars, those that are children of the same parent, as in Fig. 5.17, will each contribute to the Markov sum Eq. (5.60). Thus, if there are k order-isomorphic children $\widetilde{C}_{n+1}^{j_s(i)}$, $s \in [1, \ldots, k]$ for a given parent $\widetilde{C}_n^i$, then $\mu(\widetilde{C}_{n+1}^{j_s(i)}) = \mu_k$ for all s, which means μ_k appears with multiplicity k in the Markovian sum Eq. (5.60). In this sense, though there is covariance, order-isomorphisms are not entirely "gauge" in these models.

These three conditions on the transition probabilities define (along with internal temporality) the Rideout-Sorkin classical sequential growth (CSG) models, which as we will now see, simplify the possible growth dynamics drastically, resulting in a single, independent coupling constant at every stage [62].

Exercise 5.13 Consider a parent $\widetilde{C}_n^i$ and its children $\widetilde{C}_{n+1}^{j(i)}$. Let $\widetilde{C}_{n+1}^g$, $\widetilde{C}_{n+1}^t$ denote the **gregarious child** and the **timid child**, respectively. For all except the timid child, the transition $\widetilde{C}_n^i \to \widetilde{C}_{n+1}^{j(i)}$ has a non-trivial spectator set $S(j(i)) \subset \widetilde{C}_n^i$ and hence has non-trivial intersection with the spectator set $S^g \equiv \widetilde{C}_n^i$ of the gregarious child. Using the Bell causality condition and the Markovian sum rule show that every $p_n(j(i)) \equiv \mathbb{P}(\widetilde{C}_n^i \to \widetilde{C}_{n+1}^{j(i)})$ can be expressed in terms of $q_n \equiv \mathbb{P}(\widetilde{C}_n^i \to \widetilde{C}_{n+1}^g)$ (See [62].)

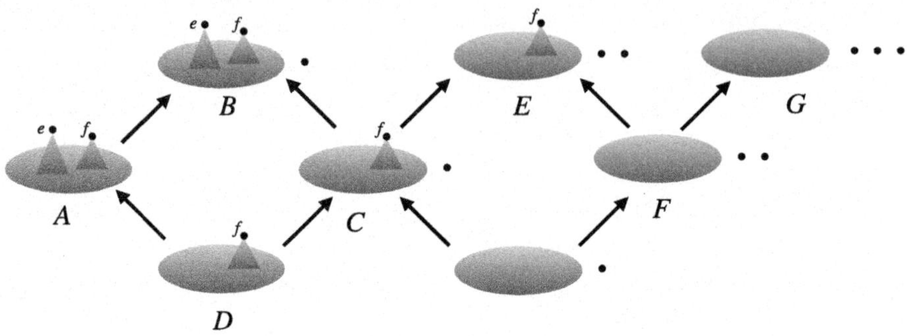

Fig. 5.22 A proof using Bell Causality and General Covariance to show that the gregarious transition is always given by q_n

Next, we prove that q_n defined in Ex. 5.13 depends only on the cardinality of its parent and hence *every* transition at stage n can be expressed in terms of q_n. For this we need a subset of unlabelled Poscau $\mathcal{P}$ (see Fig. 7 of [62] which we reproduce here in Fig. 5.22.) We start with an n-element causal set A which has a maximal element e and let B denote its gregarious child, with the transition probability $p_{AB} : A \to B$. e then remains a maximal element in B. Let D denote the parent of A which does not contain the element e, and let $p_{DA} : D \to A$. If C is the gregarious child of D, with $p_{DC} : D \to C$, then B can also be produced from C (which does not contain the element e) with probability $p_{CB} : C \to B$. This completes the diamond $\{D, A, C, B\}$. From general covariance

$$p_{DA} p_{AB} = p_{DC} p_{CB}. \tag{5.63}$$

Next, extend this causal diamond in $\mathcal{P}$ to include an adjoining diamond which shares the node C. Let E be the gregarious child of C and let the transition probability be p_{CE}. Again, wlog we can remove an element f that is maximal but not minimal in E (and therefore also maximal in C) to get the n-element causal set F, with the transition probability p_{FE}. Finally, we have the gregarious transition from F to G with transition probability p_{FG}. The second diamond is therefore completed by the causal set H obtained from C by removing the maximal element f.

The two pairs of transitions (a) $D \to A, D \to C$ and (b) $C \to B, C \to E$ are Bell-causality related. In (a) the common spectator set is the spectator set $S_a \equiv S(e)$ of the transition $D \to A$, i.e., those elements that are unrelated to the new element e, while in (b) the common spectator set is $S_b \equiv S(e) \sqcup e'$ where e' is the new gregarious element in the transition $D \to C$. Since S_a is obtained from S_b by removing e', we see that

$$\frac{p_{CE}}{p_{CB}} = \frac{p_{DC}}{p_{DA}}. \tag{5.64}$$

Equation (5.63) and (5.64) then imply that

$$p_{AB} = p_{CE} \tag{5.65}$$

In other words, the gregarious transitions $q_n(A) \equiv p_{AB} = p_{CE} \equiv q_n(C)$, and hence are all equal. We can then proceed to the adjoining diamond $HCEF$, and similarly show that $q_n(F) \equiv p_{FG} = q_n(A) = q_n(C)$. Starting with A we have chosen C in a particular way, i.e., by replacing one of the maximal elements e of A by the gregarious element e', and similarly, starting with C we have again removed one of its maximal elements which is not a minimal element f and replaced it with a gregarious element f' to get F. It is then easy to see that proceeding this way, we will ultimately reach the antichain a_n and the gregarious transition $\mathrm{a}_n \to \mathrm{a}_{n+1}$. If q_n is the associated transition probability, we have shown that starting from *any* A, $q_n(A) = q_n$.

Combined with the fact that every transition in a family is related to the gregarious transition in that family, this means that every transition at stage n can be determined from the single transition probability q_n.

One of the simplest growth models is **transitive percolation**, where $q_n = q^n$ which means that there is a single parameter q, $0 < q < 1$ that governs the entire growth process.

Exercise 5.14 Show that for transitive percolation, for any transition $\widetilde{C}_n^i \to \widetilde{C}_{n+1}^{j(i)}$ the transition probability α_n is given by

$$\alpha_n = (1-q)^m q^{n-\varphi} \tag{5.66}$$

where m is the cardinality of the set of maximal elements in the precursor set of the new element e_{n+1} and φ is the cardinality of the full precursor set.

Using this one can show [62] that for a general growth dynamics, for a generic transition at stage n, $\widetilde{C}_n^i \to \widetilde{C}_{n+1}^{j(i)}$, the transition probability is given by

$$\mathbb{P}(\widetilde{C}_n^i \to \widetilde{C}_{n+1}^{j(i)}) = \sum_{k=0}^{m} (-)^k \binom{m}{k} \frac{q_n}{q_{\varpi-k}}, \tag{5.67}$$

where ϖ is the cardinality of the precursor set, and m denotes the number of maximal elements in the precursor set. Alternatively, one can use the coupling constants t_n [62],

$$t_n = \sum_{k=0}^{n} (-)^{n-k} \binom{n}{k} \frac{1}{q_k}. \tag{5.68}$$

in terms of which the transition probabilities are

$$\mathbb{P}(\widetilde{C}_n^i \to \widetilde{C}_{n+1}^{j(i)}) = \frac{\lambda(\varpi, m)}{\lambda(n, 0)}, \quad \text{where } \lambda(a, b) = \sum_{k=b}^{a} \binom{a-b}{k-b} t_k. \tag{5.69}$$

Using these transition probabilities, we can construct a probability measure on $\mathfrak{A}$ as follows. For every node $\widetilde{C}_n^i \in \widetilde{\mathcal{P}}$, if $\gamma_n^i = \{\widetilde{C}_1, \widetilde{C}_2^{j_2}, \ldots \widetilde{C}_{n-1}^{j_{n-1}}, \widetilde{C}_n^{j_n=i}\}$ denotes the unique path in $\widetilde{\mathcal{P}}$ from $\widetilde{C}_1$ to $\widetilde{C}_n$,

$$\mu(\text{cyl}(\widetilde{C}_n^i)) = \prod_{s=1}^{n-1} \mathbb{P}(\widetilde{C}_s^{j_s} \to \widetilde{C}_{s+1}^{j_{s+1}}). \tag{5.70}$$

Note that the t_n depend on the q_k of *all* past stages and hence summarise the past. This will become important in describing cosmological renormalisation below.

Exercise 5.15 Show that (a) for transitive percolation $t_n = t^n$ and $t > 0$ (b) for the deterministic **dust universe** where only the antichain is generated, i.e., $q_n = 1$, show that $t_0 = 1, t_k = 0, k \geq 1$ (c) for the **forest universe** in which all transition probabilities are equal and are given by $\mathbb{P}(\widetilde{C}_n^i \to \widetilde{C}_{n+1}^{j_1}) = q_n = (1+n)^{-1}$ show that $t_0 = t_1 = 1$, $t_k = 0, k \geq 2$. What is the structure of the typical causal set that emerges in (c)?

Classical sequential growth models have been studied extensively and are an important stepping stone both conceptually and technically, to a full quantum dynamics. As classical models they are easier to study and have helped shape our understanding of the structures we should be quantising. Some of these results are summarised below.

A post p in a causal set (described above) has a particular cosmological significance, since it corresponds to a bouncing universe. The entire past history of the universe upto and including the post is contained in the information of the single element p. Posts in a causal set are covariantly defined and mark different eras, with the first post p_1 marking its past (possibly empty) era E_1, with E_2 the era between p_1 and p_2 and so on. As the universe bounces about p_i, the dynamics of the subsequent era E_{i+1} must be determined by parameters which are conditioned by the existence of p_i. The effective dynamics in E_{i+1} is then given by an effective, renormalised set of parameters, $\tilde{t}_n^{(i)}$ for $n > \pi_i$, where $\pi_i \equiv |\text{Past}(p_i)| + 1$ [64].

Since every element in E_{i+1} must have at least π_i elements to its past (i.e., the past of p_i including p_i itself, we can write down an expression analogous to Eq. (5.69) by defining the effective size of the past $\varphi^i = \varphi + \pi_i$, with

$$\tilde{\lambda}(\varphi^i, m) = \lambda(\varphi, m). \tag{5.71}$$

Exercise 5.16 Show using Eq. (5.71) that the effective renormalised coupling constants are

$$\tilde{t}_n^{(i)} = \sum_{k=0}^{\pi_i} \binom{\pi_i}{k} t_{n+k} \tag{5.72}$$

which implies that

$$\tilde{\lambda}(\varphi^i, m) = \sum_{k=m}^{\varphi^i} \binom{\varphi^i - m}{k - m} \tilde{t}_k. \tag{5.73}$$

This gives rise to a renormalisation map $M : t_n \to \tilde{t}_n$. As shown in [64], if $(t_n) \equiv (t_1, t_2, \ldots)$ is such that $t_i > 0$ then the renormalisation map gives rise to a fixed point for a sufficiently large n, which is transitive percolation dynamics $t_n = t^n$ for $t \geq 0$. M moreover has no cycles. This kind of *dynamical* renormalisation group type flow is therefore of great interest to quantum gravity and the hope is that some form of it also exists in the full quantum dynamics.

The notion of a post is easily generalised to that of a **break** [198,213] in a causal set c. This is an ordered pair of non-empty subsets (A, B) of c such that $c = A \sqcup B$ and

1. $a \in A, b \in B \Rightarrow a \prec b$
2. $B \supset \mathrm{Fut}(B)$

This means that all the minimal elements of B are above all the maximal elements of A. Such events can be used to renormalise the coupling constant at a causal set equivalent of a black hole singularity, which could generate the creation of new universes [198]. The set of causal set cosmologies with breaks are therefore also of interest because of the cosmological renormalisation mechanism. In [213] two event algebras for these cosmologies were constructed—one using cylinder sets and one using stem sets.

As the reader will notice, there are echoes of the return event in the RW in the post events as well events with breaks.

5.3.3 Quantum Sequential Growth Models

This section describes work in progress. At the point of writing this book, we have no realistic candidate quantum sequential growth models [70,71], that fit the bill even technically, i.e., without the complication of quantum interpretation.[5] What we have proved is that for a a simpler class of non-perturbative models of quantum gravity, covariant events are measurable [71]. Extending these results to a more realistic class of dynamics is challenging and an active area of research.

This section contains a framework for quantum sequential growth as well as a set of partial results. I have included as much detail as possible to allow the interested reader to make progress in the many directions that will present themselves.

Once one has decided on the kinematical structure of quantum gravity, as we have, the method of quantisation is intimately tied to (i) what suits that kinematical structure best, and (ii) whether it will be amenable, in the final analysis, to a closed-systems quantum interpretation. As we've stated earlier, the answer to (i) for causal sets is the path integral, and, for (ii) we use clues from Sorkin's **coevent** or anhomomorphic interpretation [30,31], and the associated **quantum measure**

[5] Whether it is possible at all to know one without the other is moot. The attitude here will be to push technically as much as possible and explore the space of all possible quantum dynamics.

framework, derived from a **decoherence functional**. The quantum amplitude of the form $A(\gamma) \equiv \exp(iS(\gamma)/\hbar)$ in the quantum partition function, is generalised to a decoherence function $D(\gamma, \gamma')$ over pairs of paths $\{\gamma, \gamma'\}$, which gives the (complex) quantum interference between the two paths.

For finite n-element causal sets the decoherence functional D is therefore defined over $\Omega_n \times \Omega_n$ (or $\widetilde{\Omega}_n \times \widetilde{\Omega}_n$ if we consider labelled causal sets.) while for the set of countable past finite sets $\widetilde{\Omega}_{\rm pf}$ it is defined over $\mathfrak{A} \times \mathfrak{A}$, where the event algebra $\mathfrak{A}$ is generated by cylinder sets.

We will require the **decoherence functional** to satisfy certain properties in order to recover some of the features of ordinary quantum theory (see [70].) We define it as a complex functional $D : \mathfrak{A} \times \mathfrak{A} \to \mathbb{C}$ which is

1. Positive: For all $\alpha \in \mathfrak{A}$, $D(\alpha, \alpha) \geq 0$.
2. Hermitian: For all $\alpha, \beta \in \mathfrak{A}$, $D(\alpha, \beta) = D^*(\beta, \alpha)$.
3. Finitely bi-additive: For $\alpha \in \mathfrak{A}$ and a set of m mutually disjoint $\beta_i \in \mathfrak{A}$, $D(\alpha, \sqcup_{i=1}^m \beta_i) = \sum_{i=1}^m D(\alpha, \beta_i)$. Similarly, for any $\beta \in \mathfrak{A}$ and m mutually disjoint $\alpha_i \in \mathfrak{A}$, $D(\bigsqcup_{i=1}^m \alpha_i, \beta) = \sum_{i=1}^m D(\alpha_i, \beta)$.
4. Normalised: $D(\widetilde{\Omega}_{\rm pf}, \widetilde{\Omega}_{\rm pf}) = 1$.
5. Strongly positive: For any finite collection of events $\{\alpha_i\}$ in $\mathfrak{A}$, the matrix $M_{ij} \equiv D(\alpha_i, \alpha_j)$ is positive semi-definite, i.e., it has non-negative eigenvalues.[6]

Thus, instead of the classical measure triple $(\widetilde{\Omega}_{\rm pf}, \mathfrak{A}, \mu)$ for CSG, the quantum sequential growth (QSG) on $\widetilde{\mathcal{P}}$ is defined by the triple $(\widetilde{\Omega}_{\rm pf}, \mathfrak{A}, D)$. We define the **quantum measure** $\mu_Q(\alpha) \equiv D(\alpha, \alpha)$, which is real and ≥ 0. μ_Q is however not additive on disjoint sets, but does satisfy the higher order sum rule [26]:

$$\mu_Q(\alpha \sqcup \beta \sqcup \gamma) = \mu_Q(\alpha \sqcup \beta) + \mu_Q(\alpha \sqcup \gamma) + \mu_Q(\beta \sqcup \gamma) - \mu_Q(\alpha) - \mu_Q(\beta) - \mu_Q(\gamma). \tag{5.74}$$

As noted in Sect. 5.3.1.2 the events in $\mathfrak{A}$ are not covariant; covariant events lie in the larger sigma-algebra $\widetilde{\mathfrak{S}}$ generated from $\mathfrak{A}$. To define a fully covariant quantum dynamics we will therefore need to extend D to a decoherence functional on $\widetilde{\mathfrak{S}}$. While there is no known quantum analogue of the KCH extension theorem for the decoherence functional, we can nevertheless use the Caratheodary-Hahn-Kluvnek (CHK) theorem for vector measures, by recasting D as a vector measure on $\mathfrak{A}$ which takes values in a **Histories Hilbert space** $\mathcal{H}_{\rm h}$, described as follows [69,70]. Define the space of complex valued functions on $\mathfrak{A}$

$$V(\mathfrak{A}) \equiv \{f : \mathfrak{A} \to \mathbb{C} | f^{-1}(\mathbb{C}\backslash\{0\}) \text{is finite}\}, \tag{5.75}$$

[6] The decoherence functional doesn't have to satisfy the condition of strong positivity, but this is a technically useful condition for defining covariant covariant quantum sequential growth dynamics as we will see below.

i.e., any $f \in V(\mathfrak{A})$ is non-zero only on a finite number of events $\alpha \in \mathfrak{A}$. $V(\mathfrak{A})$ has a natural **free vector space** structure over $\mathfrak{A}$

$$\begin{aligned}(f+g)(\alpha) &\equiv f(\alpha)+g(\alpha) \in V(\mathfrak{A}) \\ (\lambda f)(\alpha) &\equiv \lambda f(\alpha) \in V(\mathfrak{A}) \forall\, \lambda \in \mathbb{C},\end{aligned} \tag{5.76}$$

and is spanned by the characteristic vectors

$$\chi_\alpha(\alpha') \equiv \begin{cases} 1 \text{ if } \alpha' = \alpha \\ 0 \text{ otherwise.} \end{cases} \tag{5.77}$$

The decoherence functional provides a natural inner product on $V(\mathfrak{A})$

$$\langle u, v \rangle \equiv \sum_\alpha \sum_\beta u^*(\alpha) v(\beta) D(\alpha, \beta) \Rightarrow \langle \chi_\alpha, \chi_\beta \rangle = D(\alpha, \beta). \tag{5.78}$$

For sets of zero quantum measure, i.e., $D(\alpha, \alpha) = 0$, the characteristic vector χ_α is therefore of zero norm. Using a GNS-type procedure we can then construct the Histories Hilbert space $\mathcal{H}_\mathrm{h}$ as promised, by quotienting the set of all Cauchy sequences $\{u_i\}$ in $V(\mathfrak{A})$ by the equivalence relation

$$\{u_i\} \sim \{v_i\} \text{ if } \lim_{i\to\infty} ||u_i - v_i|| = 0. \tag{5.79}$$

Thus the vectors $[\{u_i\}]$ span the Hilbert space $\mathcal{H}_\mathrm{h}$ where

$$\begin{aligned}[\{u_i\}] + [\{v_i\}] &\equiv [\{u_i + v_i\}] \\ \lambda[\{u_i\}] &\equiv [\{\lambda u_i\}] \\ \langle [\{u_i\}], [\{v_i\}] \rangle &\equiv \lim_{i\to\infty} \langle u_i, v_i \rangle\end{aligned} \tag{5.80}$$

$\forall [\{u_i\}], [\{v_i\}] \in \mathcal{H}_\mathrm{h}, \lambda \in \mathbb{C}$. The **quantum vector measure,** $\mu_\mathbf{v} : \mathfrak{A} \to \mathcal{H}_\mathrm{h}$ is

$$\mu_\mathbf{v}(\alpha) \equiv [\chi_\alpha], \tag{5.81}$$

(where we have used $\{\chi_\alpha\} \equiv \chi_\alpha$) and is therefore finitely additive

$$\mu_\mathbf{v}(\sqcup_i \alpha_i) = \sum_i \mu_\mathbf{v}(\alpha_i), \quad \langle \mu_\mathbf{v}(\alpha), \mu_\mathbf{v}(\beta) \rangle = D(\alpha, \beta). \tag{5.82}$$

Finite additivity means that the triple $(\widetilde{\Omega}_\mathrm{pf}, \mathfrak{A}, \mu_\mathbf{v})$ satisfies the standard requirements for a measure space, and is therefore technically more useful to work with. In what follows, we will use the following notation interchangeably:

$$|\widetilde{C}^i_n\rangle \equiv \mu_\mathbf{v}(\mathrm{cyl}(\widetilde{C}^i_n)), \quad |\alpha\rangle = \mu_\mathbf{v}(\alpha). \tag{5.83}$$

Unlike the KCH theorem for classical measures, the CHK Theorem (see Theorem 5.1) states that $\mu_{\mathbf{v}}$ must satisfy certain conditions to extend to the full sigma algebra $\widetilde{\mathfrak{S}}$, i.e., not every choice of decoherence functional on finite causal sets (or equivalently the cylinder set nodes of $\widetilde{\mathcal{P}}$) gives rise to a covariant quantum measure on $\widetilde{\mathfrak{S}}$ (even if we start off with a "finite" covariance condition as in CSG.) In particular, the conditions for extension of a general vector measure in Theorem 5.1 [214], include convergence conditions for the norm of $\mu_{\mathbf{v}}$ in $\mathcal{B}$.

Before stating this theorem we will first need a few more definitions:

1. The **variation** of $\mu_{\mathbf{v}}$ is defined as

$$|\mu_{\mathbf{v}}|(\alpha) \equiv \sup_{\pi} \sum_{\alpha_i \in \pi} |||(\alpha_i)\rangle||, \quad \forall\, \alpha \in \mathfrak{A} \tag{5.84}$$

   where π is a **finite partition** of α, i.e., $\pi = \{\alpha_1, \ldots, \alpha_k\}, k < \infty, \alpha_i \cap \alpha_j = \emptyset, \forall\, i \neq j$ and $\alpha = \bigsqcup_{i=1}^{k} \alpha_i$. The measure is said to be of **bounded variation** if

$$|\mu_{\mathbf{v}}|(\Omega) < \infty. \tag{5.85}$$

2. The **semi-variation** $||\mu_{\mathbf{v}}||$ of a vector measure $\mu_{\mathbf{v}}$ is defined as

$$||\mu_{\mathbf{v}}||(\alpha) = sup\{|x^* \mu_{\mathbf{v}}|(\alpha); x^* \in \mathcal{H}^*, ||x^*|| \leq 1\}, \tag{5.86}$$

   where $\mathcal{H}^*$ is the dual space. Note that $x^*\mu_{\mathbf{v}}$ is an inner product which is itself a vector measure valued in $\mathbb{C}$. $\mu_{\mathbf{v}}$ is said to be **bounded** if $||\mu_{\mathbf{v}}||(\Omega) < \infty$.
3. $\mu_{\mathbf{v}}$ is **countably additive** if for every infinite sequence $\{\alpha_1, \ldots, \alpha_n, \ldots\}$ of pair-wise disjoint members of $\mathfrak{A}$ for which $\bigcup_i \alpha_i \in \mathfrak{A}$, $\mu_{\mathbf{v}}(\bigcup_i \alpha_i) = \sum_i \mu_{\mathbf{v}}(\alpha_i)$.
4. $\mu_{\mathbf{v}}$ is **weakly countably additive** if $x^*\mu_{\mathbf{v}}$ is countably additive for every $x^* \in \mathcal{H}^*$.
5. $\mu_{\mathbf{v}}$ is **strongly additive** if for every sequence $\{\alpha_n\}$ of pairwise disjoint element of $\mathfrak{A}$, $\sum_{n=1}^{\infty} |\alpha_n\rangle$ converges in the norm.

A restricted form of the CHK theorem states:

Theorem 5.1 Caratheodory-Hahn-Kluvnek: *Let $\mathfrak{A}$ be a field of subsets of Ω and $\mathfrak{S}_{\mathfrak{A}}$ be the σ-field generated by $\mathfrak{A}$. Then if $\mu_{\mathbf{v}}$ is a (i)* **bounded,** *(ii)* **weakly countably additive** *vector measure over $\mathfrak{A}$ then the following are equivalent.*

1. *$\exists$! countably additive extension of $\mu_{\mathbf{v}}$ to $\mathfrak{S}_{\mathfrak{A}}$.*
2. *$\mu_{\mathbf{v}}$ is (iii)* **strongly additive.**

The full statement of the theorem has a few more equivalence conditions which may be important for general quantum measures [214]. In this work, however, we will consider a more restricted class of quantum sequential growth (QSG) dynamics for which this statement of the theorem suffices. We point the interested reader to reference [214] for more detail on vector measure extension results.

A QSG dynamics on $\widetilde{\mathcal{P}}$ is the triple $(\widetilde{\Omega}_{\mathrm{pf}}, \mathfrak{A}, \mu_{\mathbf{v}})$ where $\mathfrak{A}$ is the event algebra generated by cylinder sets and $\mu_{\mathbf{v}} : \mathfrak{A} \to \mathcal{H}_{\mathrm{h}}$ satisfies the conditions

1. **Markovian:** On $\widetilde{\mathcal{P}}$ $\mu_{\mathbf{v}}$ is generated from the **transition matrices** $\widehat{O}(\widetilde{C}_n^i \to \widetilde{C}_{n+1}^{j(i)})$ where $|\widetilde{C}_{n+1}^{j(i)}\rangle = \widehat{O}(\widetilde{C}_n^i \to \widetilde{C}_{n+1}^{j(i)})|\widetilde{C}_n^i\rangle$, with

$$\sum_{j(i)} \widehat{O}(\widetilde{C}_n^i \to \widetilde{C}_{n+1}^{j(i)}) = \mathbb{I} \tag{5.87}$$

where, as before, $j(i)$ is valued in $\mathfrak{I}(i,n)$, and $\mathbb{I}$ denotes the identity operator on $\mathcal{H}_{\mathrm{h}}$. Associated with every $\widetilde{C}_n^i$ is a unique path $\gamma = \{\widetilde{C}_1, \widetilde{C}_2^{j_2}, \ldots \widetilde{C}_{n-1}^{j_{n-1}}, \widetilde{C}_n^{j_n=i}\}$ in $\mathcal{P}$ starting from $\widetilde{C}_1$ and ending at $\widetilde{C}_n^i$. Thus the quantum vector measure is

$$\mu_{\mathbf{v}}(\widetilde{C}_n^i) = |\widetilde{C}_n^i\rangle = \Pi_{s=1}^{n-1} \widehat{O}(\widetilde{C}_s^{j_s} \to \widetilde{C}_{s+1}^{j_{s+1}})|\Omega\rangle. \tag{5.88}$$

2. **Covariance:** For all order isomorphic causal sets $\widetilde{C}_n^i \sim \widetilde{C}_n^j$, $|\widetilde{C}_n^i\rangle = |\widetilde{C}_n^j\rangle$.[7]

The reader will notice the absence of the third condition that we used for CSG models, namely a quantum version of the Bell Causality condition. Currently we do not have a good enough conceptual understanding of quantum spectator independence via quantum nonlocality and the Bell inequalities [215]. For the remainder of this section we will restrict ourselves to **complex sequential growth** ($\mathbb{C}$SG) models defined by complexifying the coupling constants $\{t_i\}$ of the CSG models, i.e., $t_i \in \mathbb{C}$. Here, $\mu_{\mathbf{v}}$ is valued in $\mathbb{C}$ and hence the classical version of the Bell causality condition, Eq. (5.62) is easily generalised to a complex condition.

In $\mathbb{C}$SG the transition matrices of Eq. (5.87) simplify to complex transition amplitudes

$$A(\widetilde{C}_n^i \to \widetilde{C}_{n+1}^{j(i)}) \equiv \frac{\lambda(\varpi, m)}{\lambda(n, 0)}, \tag{5.89}$$

where

$$\lambda(a,b) = \sum_{k=b}^{a} \binom{a-b}{k-b} t_k, \quad \text{and} \quad \sum_{j(i)} A(\widetilde{C}_n^i \to \widetilde{C}_{n+1}^{j(i)}) = 1. \tag{5.90}$$

By the triangle inequality

$$\sum_{j(i)} |A(\widetilde{C}_n^i \to \widetilde{C}_{n+1}^{j(i)})| \geq 1, \tag{5.91}$$

---

[7] Note that this is stronger than the statement that their norms or quantum measures are the same.

where the equality is satisfied only when all the amplitudes are collinear to the unit real vector in $\mathbb{C}$. Since $|\Omega)\rangle = \sum_{i=1}^{I(n)} |\widetilde{C}_n^i)\rangle$ for any n and $|A(\Omega)| = 1$ from the normalisation condition, we also have the inequality

$$\sum_{i=1}^{I(n)} ||\widetilde{C}_n^i\rangle| \geq 1. \tag{5.92}$$

Quantifying the departure from collinearity will be important for the analysis below. For example, let $q_1 \equiv A(\widetilde{C}_1 \to \mathrm{a}_2)$ and $1 - q_1 \equiv \widetilde{C}_1 \to \widetilde{\mathrm{c}}_2$, where as before $\widetilde{\mathrm{c}}_n$, and a_n denote the n-chain and the n-antichain, respectively. For the vectors q_1 and $1 - q_1$ the triangle inequality implies that $|q_1| + |1 - q_1| \geq 1$. The equality is *only* satisfied when q_1 is real and non-negative.

For every node $\widetilde{C}_n$ in $\widetilde{\mathcal{P}}$ there is a unique path $\gamma = \{\widetilde{C}_1, \widetilde{C}_2^{j_2}, \ldots, \widetilde{C}_{n-1}^{j_{n-1}}, \widetilde{C}_n^{j_n}\}$ which defines the quantum vector measure

$$\mu_{\mathbf{v}}(\mathrm{cyl}(\widetilde{C}_n^i)) \equiv |\widetilde{C}_n^i\rangle \equiv \prod_{s=1}^{n-1} A(\widetilde{C}_s^{j_s} \to \widetilde{C}_{s+1}^{j_{s+1}}), \tag{5.93}$$

where the product is over transitions along the $(n + 1)$-jointed branch of $\mathcal{P}$ connecting $\widetilde{C}_1 = e_0 = \bullet$ to the node $\widetilde{C}_n^i$. We refer to this class of quantum measures with complex t_n as **complex sequential growth** ($\mathbb{C}$SG) models, with decoherence functional

$$D(\alpha, \beta) = \mu_{\mathbf{v}}^*(\alpha)\mu_{\mathbf{v}}(\beta). \tag{5.94}$$

For such a "product" decoherence functional, it is easy to show that the histories Hilbert space $\mathcal{H}_{\mathrm{h}} \simeq \mathbb{C}$. Consider two distinct non-zero norm vectors $u, v \in V$ which are also therefore vectors in $\mathcal{H}_{\mathrm{h}}$. Using Eq. (5.94) and for any $\lambda \in \mathbb{C}$

$$\| u - \lambda v \|^2 = |S_1|^2 - (\lambda S_1^* S_2 + \lambda^* S_1 S_2^*) + |\lambda|^2 |S_2|^2 = |S_1 - \lambda S_2|^2, \tag{5.95}$$

where $S_1 = \sum_{\alpha \in \mathfrak{A}} \mu_{\mathbf{v}}(\alpha) u(\alpha)$ and $S_2 = \sum_{\alpha \in \mathfrak{A}} \mu_{\mathbf{v}}(\alpha) v(\alpha)$. Choosing $\lambda = S_1/S_2$ ($S_2 \neq 0$), we find that $\| u - \lambda v \| = 0$. Hence for every pair u, v of non-zero vectors in V, there exists a $\lambda \in \mathbb{C}$ such that $\{u\} = \lambda\{v\}$ and hence $\mathcal{H}_{\mathrm{h}} \simeq \mathbb{C}$.

For this class of dynamics it was shown in [71,201]:

Corollary 5.1 *For any quantum measure triple $(\widetilde{\Omega}_{\mathrm{pf}}, \mathfrak{A}, \mu_{\mathbf{v}})$, where $\mathfrak{A}$ is the event algebra generated from finite set operations on cylinder sets and $\mu_{\mathbf{v}} : \mathfrak{A} \to \mathbb{C}$, $\mu_{\mathbf{v}}$ has a unique extension to the sigma-algebra $\widetilde{\mathfrak{S}}$ associated with $\mathfrak{A}$ iff it is of bounded variation.*

Proof Assume that $\mu_{\mathbf{v}}$ is of bounded variation. From [214] it is also strongly additive, which in turn implies that it is bounded. To prove the converse, assume that $\mu_{\mathbf{v}}$ is bounded, i.e., $|x^*\mu_{\mathbf{v}}|(\Omega) < \infty$ for all $x^* \in \mathcal{H}^*$. Let $x^* = 1 (\in \mathbb{C}^* = \mathbb{C})$ which means that $|\mu_{\mathbf{v}}|(\Omega) < \infty$. Thus bounded variation is equivalent to the conditions of boundedness and strong additivity.

For every $\alpha \in \mathfrak{A}$, moreover, there exists a smallest $n < \infty$ and a subset $S \subset \{1, \ldots, \mathfrak{N}_n\}$ such that $\alpha = \bigsqcup_{k\in S} \mathrm{cyl}(\widetilde{C}_n^k)$. Thus, $\mu_{\mathbf{v}}$ is countably and weakly countably additive.

From Theorem 5.1 therefore, bounded variation of $\mu_{\mathbf{v}}$ is sufficient for it to extend to $\widetilde{\mathfrak{S}}$. Since the complex measure on any σ-algebra is of bounded variation (see Theorem 6.4 in [216]), it is also necessary. □

Thus bounded variation is both a necessary and a sufficient condition for a complex measure to extend to the full sigma algebra $\widetilde{\mathfrak{S}}$.

We begin with the simplest model, **complex percolation** ($\mathbb{C}$P), which is complexified transitive percolation. This is described by a single complex number q, which is the amplitude for a gregarious transition in $\widetilde{\mathcal{P}}$.

Claim For any $n \geq 2$, the equality in (5.92) is satisfied only when the parameter q is real. □

Proof For $n = 2$, if $|1-q| + |q| = 1$, then q is non-negative and real. Next, consider the two $n > 2$ element causal sets, (a) the n-chain $\widetilde{c}_n$ and (b) $\widetilde{C}_n^{\mathrm{V}}$, an $n-2$ chain topped with a "V", i.e. with e_n, e_{n-1} to the immediate future of the maximal element e_{n-2} of $\widetilde{c}_{n-2}$, (see Fig. 5.23.) Their quantum measures are $|\widetilde{c}_n\rangle = (1-q)^{n-1}|\Omega\rangle$, and $|\widetilde{C}_n^{\mathrm{V}}\rangle = (1-q)^{n-1} q|\Omega\rangle$, respectively. Requiring that these two vectors are collinear means that q must be non-negative and real. □

Claim If $|q| > 1$ or $|1-q| > 1$, the quantum vector measure is not of bounded variation. □

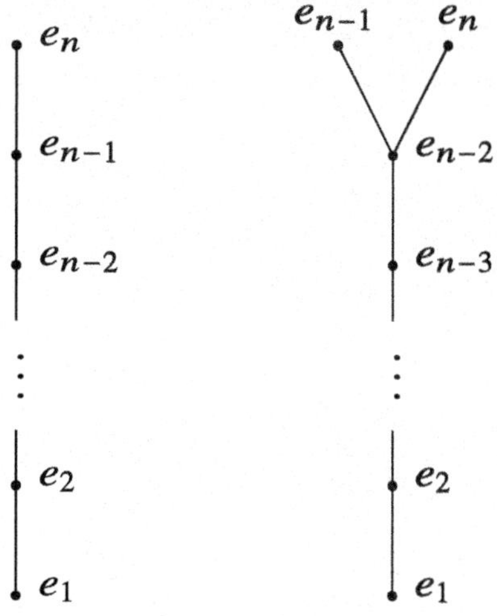

Fig. 5.23 Hasse diagrams for the chain $\widetilde{c}_n$ and "Y" poset $\widetilde{C}_n^{\mathrm{V}}$

Proof Consider any finite partition of $\widetilde{\Omega}_{\text{pf}}$ which contains $\text{cyl}(\widetilde{a}_n)$. Since $|\widetilde{a}_n\rangle = q^{n(n-1)}$, $|\mu_{\text{v}}|(\Omega) > |q|^{n(n-1)}$ if $1-q \neq 0$. Thus, if $|q| > 1$, $|\mu_{\text{v}}|(\Omega)$ grows without bound with n and hence μ_{v} is not of bounded variation. Similarly, for any finite partition of Ω which contains $\text{cyl}(\widetilde{c}_n)$, since $|\widetilde{c}_n\rangle = (1-q)^{n-1}$, the total variation $|\mu_{\text{v}}|(\Omega) > |(1-q)^{n-1}|$ as long as $q \neq 0$. Thus for $|1-q| > 1$ this grows without bound with n and hence μ_{v} is not of bounded variation. □

Using the natural partitions of a generic node $\text{cyl}(\widetilde{C}_n^i)$ in $\widetilde{\mathcal{P}}$ obtained by all its stage $n+m$ descendants it was shown in [70] that even when $|q|, |1-q| \leq 1$, $\mathbb{C}$P is not of bounded variation. Namely, we have the no-go theorem:

Lemma 5.1 *The quantum vector measure of complex percolation is not of bounded variation when the parameter q is not real.*

Thus, the simplest non-trivial strictly complex model of QSG is not covariant, in the sense that we have defined above. Note that for real non-negative q the dynamics (**real quantum percolation**) is *distinct* from transitive percolation since the quantum measure $D(\alpha, \alpha) = |\mu_{\text{v}}(\alpha)|^2$ and is therefore non-additive. However, the complex, or amplitude measure μ_{v} is additive and non-negative, which means the classical CHK extension theorem holds. This is however a trivial quantum dynamics with $D(\alpha, \beta) = \mu_{\text{v}}(\alpha)\mu_{\text{v}}(\beta)$ and mimics CSG closely.

Thankfully, this is not a death-knell for $\mathbb{C}$SG models – several families of nontrivial $\mathbb{C}$SG dynamics were shown in [71] to satisfy bounded variation. A simple example is if $t_k \neq 0$ for any $k > 0$ and $t_{k'} = 0, \forall k' \neq k, k' > 0$. However simple, they warrant further investigation.

These results for the $\mathbb{C}$SG models are currently all we know about QSG. In order to generalise to higher dimensional $\mathcal{H}_{\text{h}}$ the simple factorisation of the decoherence functional should be replaced by some more complicated rule. In unitary quantum systems the decoherence functional comes dressed with a delta function that ties the ends of the two paths together at the final time. Is there an analogue of this in QSG? Even if we had a few candidate decoherence functionals to work with it would doubtless improve our understanding of QSG models.

5.4 Summary

In this chapter we have presented various candidate dynamics for causal sets, starting from continuum inspired ones that use the discrete path integral, to the more principles based QSG dynamics in the framework of quantum measure theory. Ultimately these models must be guided by the physical questions we wish to answer.

We showed an important non-perturbative result in the discrete path sum of CST, namely that the overwhelmingly dominant entropic contribution to the path sum is suppressed by the BDG action. This simultaneously gives a lower bound on the discreteness scale, which is of order the Planck scale. An important step is to extend these results to other contributions, especially the continuum.

Next, we showed that an analytic continuation of the path sum can be studied using MCMC numerical techniques. Using the order invariants or covariant observables of Chap. 3 as order parameters, we saw that a class of dimensionally restricted causal sets exhibit a first order phase transition. Using the path sum to construct the Hartle-Hawking wavefunction we then saw that that in $d = 2$, that it is well peaked, with the peak shifting sharply at the phase transition. If this analysis survives in higher dimensions, it may play a role in early universe physics.

In the QSG models, physical questions are framed using covariant events some of which, like the post event, are clearly of cosmological relevance [70,71]. These models are motivated by the well studied CSG models [62–64,66,208,217]. An important existence result shows that although the quantum measure is non-additive, every covariant event is quantum measurable [71].

Much work remains in extracting more predictive results from CST dynamics. In this chapter we have laid the groundwork for framing the ongoing and future work in CST dynamics.

Phenomenology and Future Directions

6

Abstract

Quantum gravity phenomenology opens up new ways in which we can test our esoteric theories of spacetime. Though small, these effects can add up over the age of our universe leading to potentially measurable effects. An important prediction from CST is Sorkin's accurate order of magnitude estimate of the fluctuation in the cosmological constant, with the prediction preceding the observation by a decade. Another class of predictions comes from the swerves model of particle propagation where the discreteness leads to a slow diffusion in momentum space with potentially observable consequences. CST phenomenology is a rich subject and this chapter serves as a sampler.

6.1 Introduction

Research in quantum gravity phenomenology has seen a sharp rise in the past couple of decades. Although Planck energies are unreachable in any current human experiment, the hope is to look for phenomenological signatures from the early universe, which can leak into accessible energy regimes over sufficiently long time scales (the so called "secular processes".) This has given rise to tests of Lorentz invariance which in turn have placed strong constraints on Lorentz violation (see [73,97]). At these scales, CST is therefore on the right side of known physics since, as we have seen in Chap. 2, continuumlike causal sets do not violate local Lorentz invariance despite discreteness.

A striking result from CST is Sorkin's 1987 prediction [48,74,75] for a small but non-zero fluctuation in the cosmological constant, $\Delta\Lambda \sim 10^{-120}$ which is of the same order of magnitude obtained from subsequent observations of Λ [218–220]. This prediction gives causal set quantum gravity its strongest fillip and is one of its most convincing phenomenological motivations. Such startling accuracy from a theory of quantum gravity is indeed rare.

S. Surya, *The Causal Set Approach to Quantum Gravity*, Lecture Notes in Physics 1036,
https://doi.org/10.1007/978-3-031-84420-1_6

As a phenomenological model, of course some assumptions must be made. This is not a prediction from the quantum dynamics of the theory, but results from its kinematical structure, or what we have referred to as its first quantisation. There have subsequently been "post"-dictions, using the same copy-cat mathematical ingredients, but these are poorly motivated. As we have seen in CST, the Poisson distribution plays a key role in the continuum approximation, and hence the argument of [74] is quintessentially a CST one.

In Sorkin's original argument, fluctuations in Λ are generated kinematically [48, 74,75] and are related directly to the size of the present universe. Subsequent to the observational result, a dynamical stochastic model of Λ was constructed based on the original ideas [76]. More recently, these models have been studied extensively with explicit comparisons made with the rich observational data that is now available [78–80].

In a continuumlike causal set, the random discreteness means that the trajectory of particles will necessary "swerve", since there are no "straight" linked-paths available in a Poisson sprinkling! This is captured by the **swerves** model of particle propagation in CST [96,221,222]. The swerves, or momentum kicks, lead to a diffusion in momentum space which, over time, mean that very high energy events becomes non negligible. Therefore such a diffusion though small, can accumulate over cosmic time scales, leading to a "heating up" from extremely energetic particles. This in turn puts limits on the associated diffusion constant [223]. Recent work on the heating up of dark matter by swerves compares well with observational constraints and claims to resolve certain cosmological tensions [224].

In this last chapter we start with a very brief review of these results in CST phenomenology in Sect. 6.2. A full discussion of these results is outside the scope of this book since it involves careful comparison with observational data. We therefore urge the reader to look at the literature for details. In Sect. 6.3 we discuss some of the open questions and suggest future directions of research in CST.

6.2 CST Phenomenology

In the 1987 ICGC conference in Goa, India, soon after the first CST paper was published [2], Sorkin showed, using a simple "back-of-the-envelope" calculation, that in CST $\Delta\Lambda \sim 10^{-120}$ in Planck units, if the discreteness scale is of order the Planck scale [74]. This result was forgotten at that time, since it was assumed that Λ is "naturally" zero, despite the standard quantum field theory prediction that it should be of order ~ 1, which is clearly unphysical. About a decade later, observation showed a value of $\sim 10^{-122}$ in Planck units, which makes the Sorkin prediction startlingly accurate! It however remains one of the most undervalued predictions of our time.

The argument is very simple, and from what we have learnt in this book, the reader will see that it is quintessentially a CST result. Others have argued that it is more general, but whichever way one wants to spin it, the simplicity and accuracy of the CST result cannot be ignored.

In CST a finite region of our universe V corresponds to a finite number of causal set elements n. If we consider our present universe, within the Hubble time, this again has a finite spacetime volume within the past light cone. Thus, it makes sense to fix the number of fundamental spacetime events or causal set elements and to consider the fluctuations about this value arising from the random discreteness. In other words, the volume of our past universe V is subject to fluctuations ΔV. How should we interpret these fluctuations semi-classically?

In the Einstein-Hilbert action the cosmological constant term is

$$\frac{1}{8\pi}\Lambda\int dV, \tag{6.1}$$

which for a finite spacetime region, like our Hubble-limited past, is a constant. If we impose the constraint of a fixed total past volume, we see that Λ becomes a Lagrange multiplier. It is well known that this unimodular modification of general relativity gives rise to the Einstein equations with a cosmological constant [48,49]. Λ and V are thus canonically conjugate variables, which on quantisation satisfy the uncertainty relation

$$\Delta V \Delta\Lambda \sim 1. \tag{6.2}$$

Thus, on very general grounds, within the framework of unimodular gravity, we expect fluctuations in Λ

$$\Delta\Lambda \sim (\Delta V)^{-1}. \tag{6.3}$$

In general, there is no way to guess the value of ΔV.

However, CST kinematics not only tells us *why* V must fluctuate but also by *how* much! In CST the number n of elements is fundamental and hence fixed, which means that while $V \sim n/\rho$, there are Poisson fluctuations

$$\Delta V \sim \sqrt{V}. \tag{6.4}$$

For our visible universe $V = 10^{240}\ell_p^4$. Inserting into Eq. (6.3) we find that

$$\Delta\Lambda \sim 10^{-120}\ell_p^{-2}, \tag{6.5}$$

which is Sorkin's 1987 CST prediction.

Moreover, *assuming* $\langle\Lambda\rangle = 0$, we see that

$$\Lambda \sim \frac{1}{\sqrt{V}} \sim H^2, \tag{6.6}$$

where H is the Hubble constant. Thus under the assumption that the mean value of Λ is zero, Λ therefore always tracks the critical density, i.e., it is **ever-present**.

The role of the Planck length ℓ_p is clear—if the discreteness scale is chosen to be orders of magnitude away from ℓ_p, it changes the value of $\Delta\Lambda$ significantly. This is the case for spacetimes with large extra dimensions [225], and provides an argument against such models.

In [226] an analogy was made between the fluctuating Λ in cosmology and the surface tension T of a fluid membrane. Here, the presence of a fundamental atomicity makes the mean value of T zero, since the removal or addition of the membrane molecules does not change the surface tension. Proving $\Lambda = 0$ in CST is however considerably more difficult and remains an open question.

After the observational confirmation of dark energy of the same order of magnitude as Sorkin's prediction, a classical dynamical model for generating fluctuations of Λ was developed in [76]. A fluctuating Λ means that both of the Friedman equations in cosmology cannot simultaneously be satisfied, but at most only a linear combination [77]. For example, one can begin with the Friedmann equation

$$3\left(\frac{\dot{a}}{a}\right)^2 = \rho + \rho_\Lambda, \tag{6.7}$$

with

$$\rho_\Lambda = \Lambda, \qquad p_\Lambda = -\Lambda - \dot{\Lambda}/3H. \tag{6.8}$$

Λ is modeled in [76] as a stochastic function of the past volume V, with fluctuations given by Sorkin's CST result

$$\Delta\Lambda \sim \frac{1}{\sqrt{V}}. \tag{6.9}$$

Since Λ can be viewed as the action S per unit volume, in CST this means that every element contributes $\pm\hbar$ to S. As the universe grows with the addition of a new element, S increases by $\hbar$ and hence

$$S = \sum_{elements} \pm\hbar \Rightarrow S/\hbar \sim \pm\sqrt{N} \sim \pm\sqrt{V/\ell^4} \Rightarrow \Lambda \sim \pm\frac{\hbar/\ell^2}{\sqrt{V}}, \tag{6.10}$$

where $\ell = \sqrt{\hbar\kappa}$. This leads to the integro-differential equations

$$\frac{da}{a} = \sqrt{\frac{\rho + \Lambda}{3}}d\tau,$$
$$V d\Lambda = V d(S/V) = dS - \Lambda\dot{V}d\tau$$

where

$$V(\tau) = \frac{4\pi}{3}\int_0^t dt' a(t')^3 \left(\int_0^{t'} dt'' \frac{1}{a(t'')}\right)^3 \tag{6.11}$$

is the volume of the causal past. This defines a stochastic evolution, starting from the scale factor a_i (scale factor), N_i, V_i, S_i and Λ_i at step i which is updated using the discrete form of the Friedmann equation

$$a_{i+1} = a_i + a_i\sqrt{\frac{\rho + \Lambda}{3}}(\tau_{i+1} - \tau_i). \tag{6.12}$$

$V_{i+1} = V(\tau)$ can be calculated from Eq. (6.11) and therefore $N_{i+1} = V_{i+1}/\ell^4$. In the model of [76] the new updated action is taken to be

$$S_{i+1} = S_i + \alpha\, \xi \sqrt{N_{i+1} - N_i}, \tag{6.13}$$

where ξ is a Gaussian random variable, with $\Delta\xi = 1$, α being a tuneable free parameter which controls the magnitude of the fluctuations. Using this, the new value of the cosmological constant is

$$\Lambda_{i+1} = S_{i+1}/V_{i+1}, \tag{6.14}$$

where it is assumed that $S_0 = 0$. Simulations moreover confirm that Λ is **everpresent** and tracks the energy density of the universe. This model has been studied and developed further in recent work [79,80], giving a much clearer understanding of the constraints on the parameters, and its cosmological consequences.

As in most cosmological models one has made the simplifying assumption of homogeneity. While small inhomogeneities are expected, modeling this in standard cosmology is challenging. For the everpresent model the authors of [227,228], modelled inhomogeneities by allowing $\Lambda(x^\mu)$, with the fluctuations $\Delta\Lambda(x)$ being dependent only on the value of Λ at $y \in J^-(x)$. In the *standard* cosmology this would mean that well separated patches in the CMB sky would contain uncorrelated fluctuations in Ω_Λ. Given a strong observational constraints of $\alpha < 10^{-6}$ the authors of [227,228] argue that this is insufficient to account for Λ. While the model for inhomogeneity may be too simplistic, this remains an open question and needs to be addressed in the future.

We end this section with a brief discussion on swerves [96,221,222,229]. In any continuumlike causal set, no chain or linked-path is straight with respect to the continuum embedding, even in $\mathbb{M}^d$. Thus, a particle propagating on a background causal set must suffer little kicks that make it "swerve". This change in direction means that there is a stochastic change in the momentum.

Let us assume that the center of mass of the particle can be localised at the element e_n in the causal set and then moves to the element e_{n+1} at the next (macroscopic) time step. We can take the effective momentum $\mathbf{p}_{n+1}$ to be proportional to the spacetime vector from e_n to e_{n+1} in the embedding. In the absence of external forces, at the next time step, the center of mass should be at some $e_{n+2} \succ e_{n+1}$ such that $\mathbf{p}_{n+2}$ is as close to $\mathbf{p}_{n+1}$ in both magnitude and direction, but also such that the proper time elapsed is less than some forgetting time τ_f.

In the continuum such a process manifests as a proper time diffusion in phase space $\mathbb{H}^3 \times \mathbb{M}^4$, where $\mathbb{H}^3$ is the momentum mass shell. As shown in [96] the probability distribution is a function on phase space and satisfies a Lorentz invariant diffusion equation [96,230]

$$\frac{\partial}{\partial\tau}\rho = k\Delta_p^2\rho - \frac{1}{mc^2} p^\mu \frac{\partial}{\partial x^\mu}\rho, \tag{6.15}$$

where the diffusion constant k is a parameter which depends on τ_f.

Because this is a diffusion in momentum, its clear that the swerves do not conserve energy. However, this does not contradict the current astrophysical bounds on Lorentz violation [73,97]. For protons, laboratory and astrophysical data put a very stringent a limit on k [223]. However, the possibility that swerves can explain the presence of ultra high energy neutrinos is still open. Upper limits on k mean that diffusion times are long and hence for these ultra high energies to be likely, the origin of the neutrinos must be cosmological. Current observational data has not ruled this out and if confirmed, this would be an important milestone in CST phenomenology. Recent work on the swerving of dark matter particles [224] shows that there is a spontaneous heating at late times which suppresses the matter power spectrum in an appropriate range, and thus can resolve certain cosmological tensions.

There are other aspects of CST phenomenology which this book does not cover but are very interesting [231–236]. An example is a proposal for off-shell dark matter [182,237] which arises due to nonlocality but which decouples at later times. We have given a very brief glimpse into the rich area of CST phenomenology and refer the interested reader to the many references given in the text.

6.3 Future Directions

As we have seen, CST is a well thought out and mature formulation of quantum gravity. From its roots in Lorentzian geometry, it posits a very specific, covariant fundamental spacetime discreteness and provides a new conception of the "essence" of spacetime (see Chaps. 1 and 2.) In Chap. 3 we saw how geometric and topological features of a spacetime can be reconstructed from the purely order theoretic features of the underlying ensemble of random causal sets. In Chap. 4 the free quantum scalar field theory was defined on certain continuumlike causal sets with the discreteness scale providing a covariant UV cut-off. In Chap. 5 various types of covariant quantum dynamics on causal sets were presented using the discrete version of the path integral. Finally, in this chapter we touched upon CST phenomenology (sadly too briefly), a topic that deserves its own book!

In this book I have attempted to present the basic aspects of CST using examples wherever possible, sometimes pointing to open questions in the field, and interspersing exercises where appropriate. The focus is on pedagogy and to make the exposition as clear as possible, even if it included repetition in places (reminders are always useful I find.) Given the prerequisite background of general relativity and Quantum Field Theory, I hope the book will help the interested reader understand the field and equip them for research in CST. I emphasise—the book is not intended as a complete nor fully up to date picture of CST, so we urge the reader to peruse the references which I hope are as complete as possible.

CST has many successes, but as in all approaches these are tempered by the fact that there is much left to be done. This book has hopefully described the limits of our knowledge about CST, and hence presented the frontiers of research in the field. There are many such frontiers, and I will list a very small number of these below, which also reflect my own biases. As the reader would have noticed, many smaller open questions have already been posed in the text, so I will not repeat these here.

1. Kinematics: A proof of the Hauptvermutung. Importantly, how do we define what is meant by two spacetimes being close, up to the covariant scale ρ^{-1}?
2. Kinematics: How does one define vector and tensor fields on causal sets, as well as abelian and non-abelian gauge fields? Can geometry be fully scalarised?
3. Quantum Field Theory: What is the effect of the non-scaling behaviour in the SJ spectrum on the RG flow in standard Quantum Field Theory?
4. Quantum Field Theory: Does the SSEE volume law say something about the nonlocality of causal sets and does it suggest that the area-entropy relationship is only emergent and not fundamental?
5. Dynamics: Is it possible to show that dimension $d = 4$ is emergent in the CST partition function?
6. Dynamics: Does the suppression of entropy from layered posets provide clues to constructing a well motivated double path integral for CST?
7. Dynamics: Does the first order phase transition in the state sum model lead to potential predictions for the early universe?
8. Dynamics: Further explorations on covariant quantum sequential growth, including quantum measures valued in higher dimensional histories Hilbert spaces.
9. Phenomenology: Spatially varying ever present Lambda models and whether they resolve or exacerbate the Hubble tensions we see from observations.

It is my hope that the coming decade will lead to new and significant results in CST, and quantum gravity more generally. Importantly, CST has brought, and will continue to bring, a richer flavour to the broad activity of quantum gravity rooted as it is in the undeniable fact that our universe is fundamentally Lorentzian in character.

Examples of Causal Sets

A

Abstract

In this appendix we provide a list of different types of causal sets, some of which can be used as alternatives for the excercises.

An explicit enumeration of the set Ω_n of unlabelled posets is known upto $n = 16$ [238] and the values of $|\Omega_n|$ are available as Sequence A000112 of "The On-Line Encyclopedia of Integer Sequences" [239]. The enumeration of the set $\widetilde{\Omega}_n$ of the set of labelled posets is known upto $n = 18$ and the values of $|\widetilde{\Omega}_n|$ are available as the Sequence A001035 in the same reference [240]. Figure A.1a shows $|\Omega_n|$ and $|\widetilde{\Omega}_n|$ as a function of n.

As we have discussed earlier in the text, the ratio $R_n = |\widetilde{\Omega}_n|/|\Omega_n|$ can be compared to $n!$, the number of relabellings on n elements, as shown in Fig. A.1b. As n becomes larger, the two approach each other with the "error" $(n! - R_n)/R_n$ in Fig. A.1c dipping to 0.123611 at $n = 16$. The difference is caused by the possibility of non-trivial **automorphisms** in a labelled causal set. Let $\widetilde{C} \equiv l(C)$ be a labelling of C and consider a relabelling $L : \widetilde{C} \to \widetilde{C}'$ of C. If $\widetilde{C}' = \widetilde{C}$, then L is said to be an automorphism of $\widetilde{C}$. Such automorphisms exist whenever there are **indistinguishable pairs** $e, e' \in C$, i.e., for which $Past(e) = Past(e')$ and $Fut(e) = Fut(e')$. Starting with any labeling $\widetilde{C} \equiv l(C)$ the switching of labels of indistinguishable pairs, $l(e) \leftrightarrow l(e')$ gives back the same labelled causal set $\widetilde{C}$. Hence the number of labelings is less than $n!$—one must instead divide $n!$ by the number of automorphisms.

S. Surya, *The Causal Set Approach to Quantum Gravity*, Lecture Notes in Physics 1036,
https://doi.org/10.1007/978-3-031-84420-1

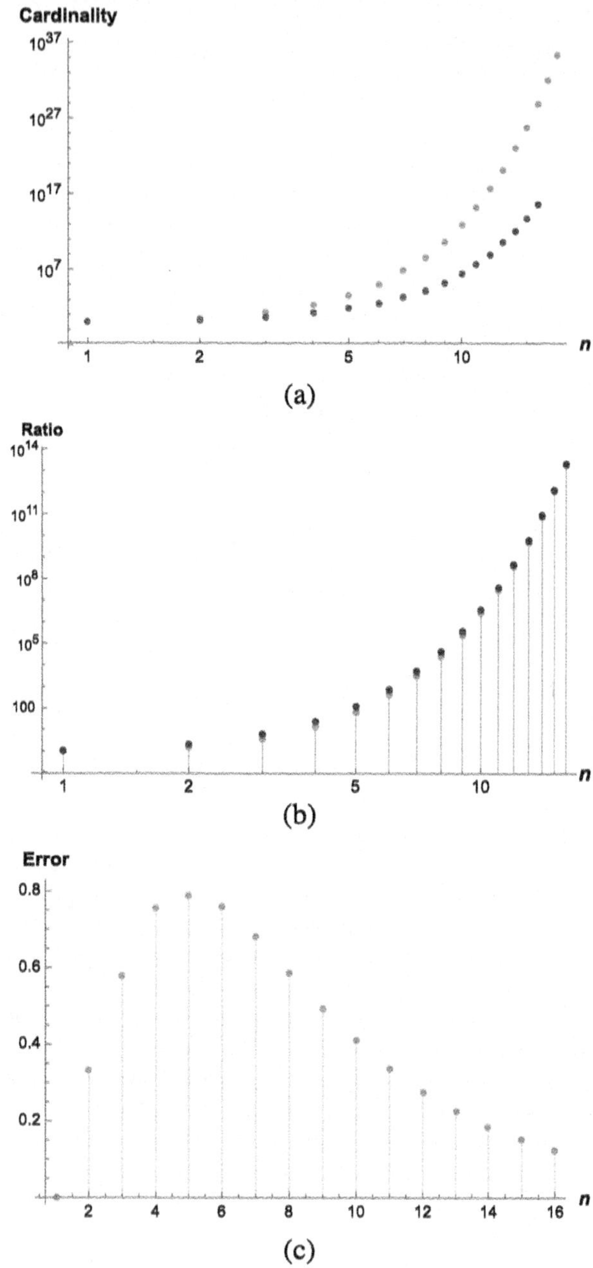

Fig. A.1 **a** A log-log plot of the cardinality of the set of causal sets as a function of the size n of the causal set, for $n = 1$ to $n = 16$. The points in dark grey are the unlabelled causal sets, i.e., $|\Omega_n|$, while those in light grey are the labelled causal sets $|\widetilde{\Omega}_n|$. **b** A log-log plot of the ratio of the number of labelled to unlabelled posets (dark grey) compared to $n!$ (light grey). As n becomes larger, the two graphs merge. **c** The "error" as a function of n

In [191] it was shown that

Theorem A.1 (Promel) *There exists a constant s such that for all n,*

$$|\Omega_n| \le \frac{|\widetilde{\Omega}_n|}{n!}\left(1 + \frac{s}{2^{n/4}}\right)$$

This implies that

Corollary A.1 (Promel) *Almost all partial orders are* **rigid**, *i.e., have no non-trivial automorphism.*

The following figures are examples of unlabelled causal sets that can be used to supplement the exercises in the book (Figs. A.2, A.3, A.4, A.5, A.6, A.7, A.8, A.9).

(a) c_2 (b) a_2

Fig. A.2 The set Ω_2 of $n = 2$ unlabelled posets, $|\Omega_2| = 2$ consists of **a** the chain c_2 and **b** the antichain a_2. The number of naturally labelled $n = 2$ posets $|\Omega_2^{(nl)}| = 2$ as well, since there is only one natural labelling each for both the c_2 and the a_2. The number of labelled posets on the other hand is $|\widetilde{\Omega}_2| = 3$: c_2 allows 2! relabellings but a_2, whose automorphism group is also of cardinality 2, does not admit any relabellings

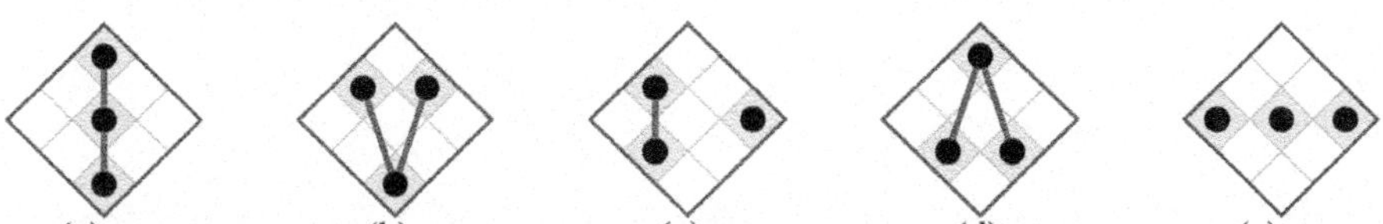

(a) c_3 (b) (c) (d) (e) a_3

Fig. A.3 The set Ω_3 of $n = 3$ unlabelled posets, has cardinality $|\Omega_3| = 5$. The posets **a**, **b**, **d** and **e** allow only one natural labelling each, while **c** admits three. Hence, $|\Omega_3^{(nl)}| = 7$. The number of *all* possible labellings is distributed as follows: **a** has $3! = 6$, **b** has $3!/2! = 3$ since the number of automorphism is 2, **c** has $3! = 6$, **d** has again $3!/2!$ since the number of automorphims is 2 and finally **e** has $3!/3! = 1$ since the number of automorphims is 3!. Thus, $|\widetilde{\Omega}_3| = 19$

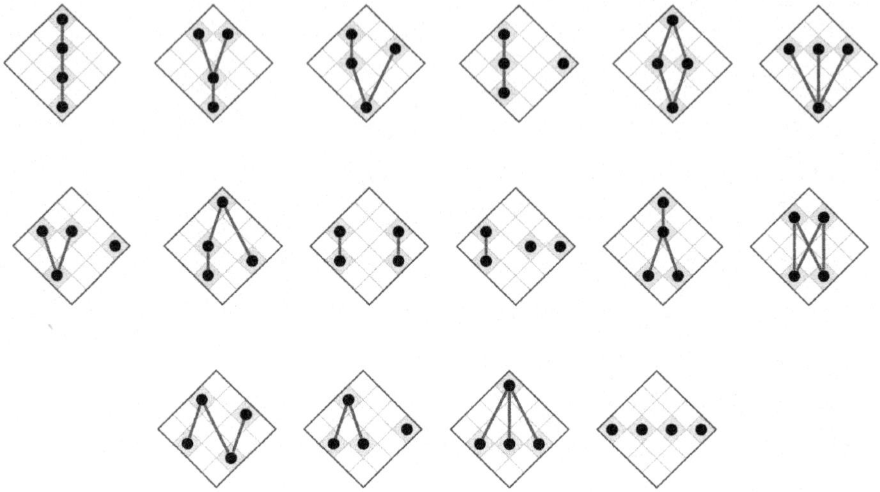

Fig. A.4 The set of $n = 4$ posets

Fig. A.5 The set of $n = 4, 6, 8$ crown posets

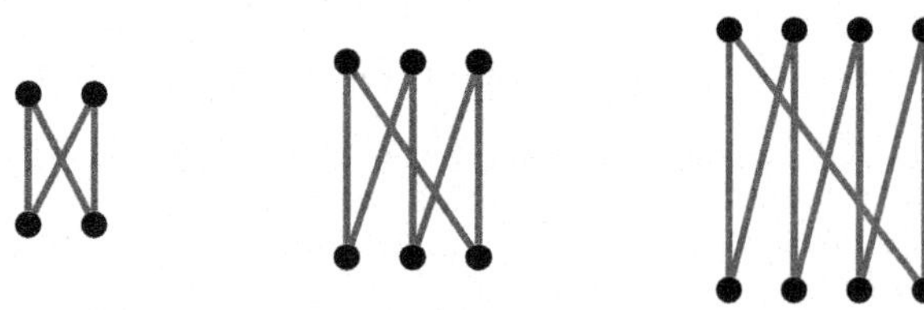

Fig. A.6 An example of a dominant $n = 16$ bilayer order with $dn^2 = \frac{1}{8}n^2$ relations

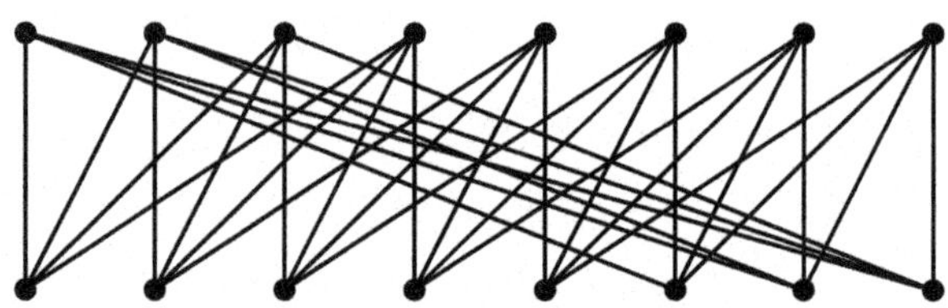

Fig. A.7 An example of an $n = 16$ KR order with $dn^2 = \frac{3}{16}n^2$ relations

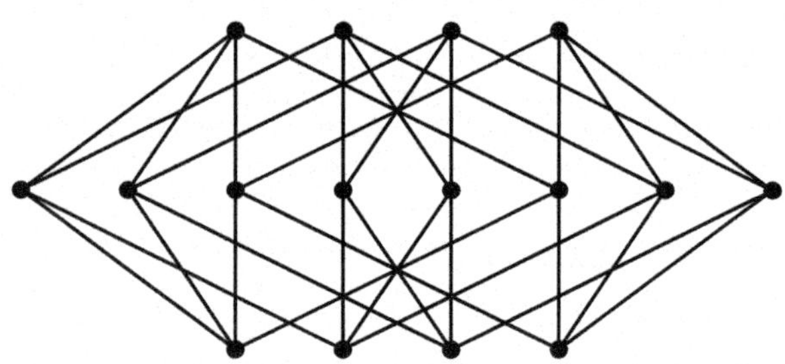

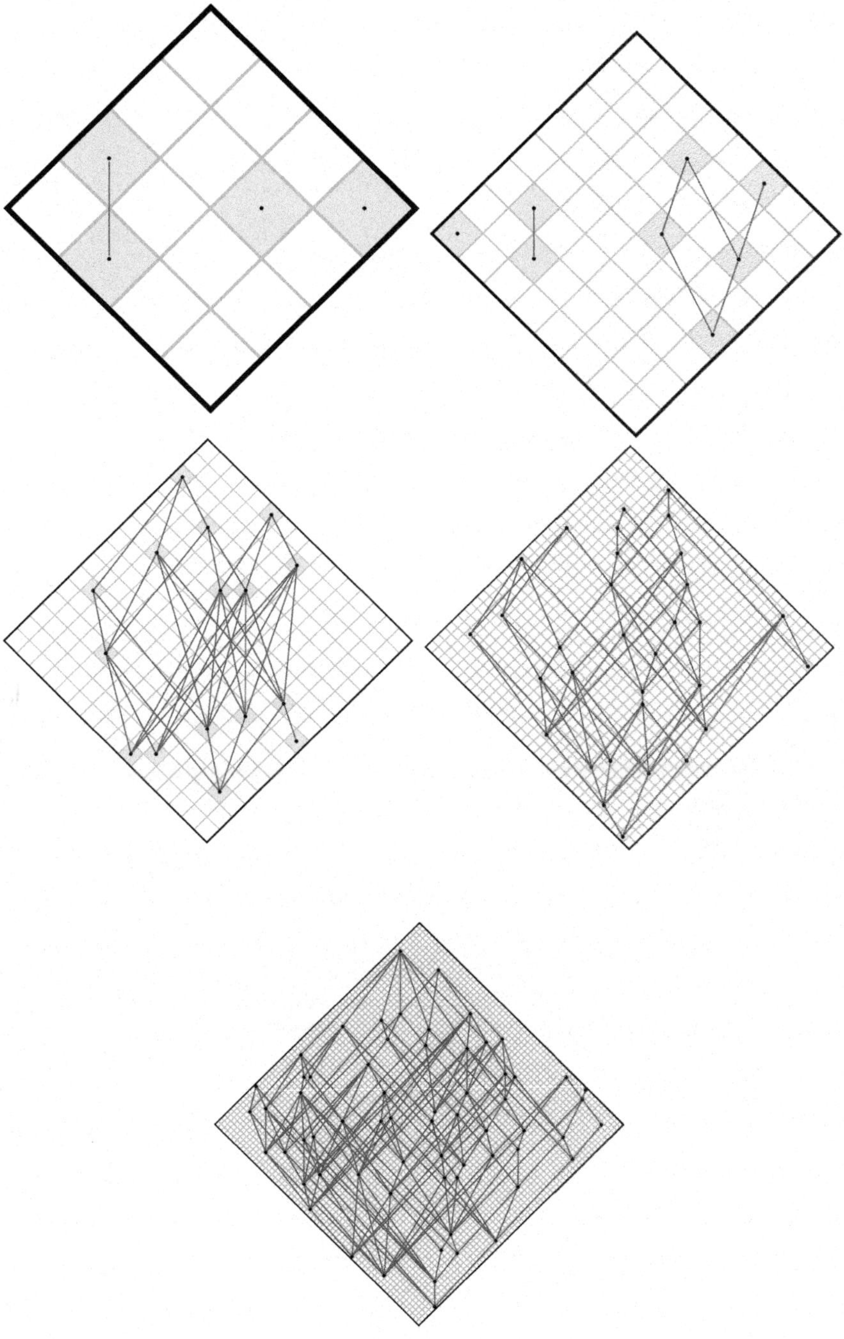

Fig. A.8 Random 2d orders of size $n = 4, 8, 16, 32, 64$. As the density of sprinkling increases, the connectivity does as well, and hence captures more of the continuum geometry

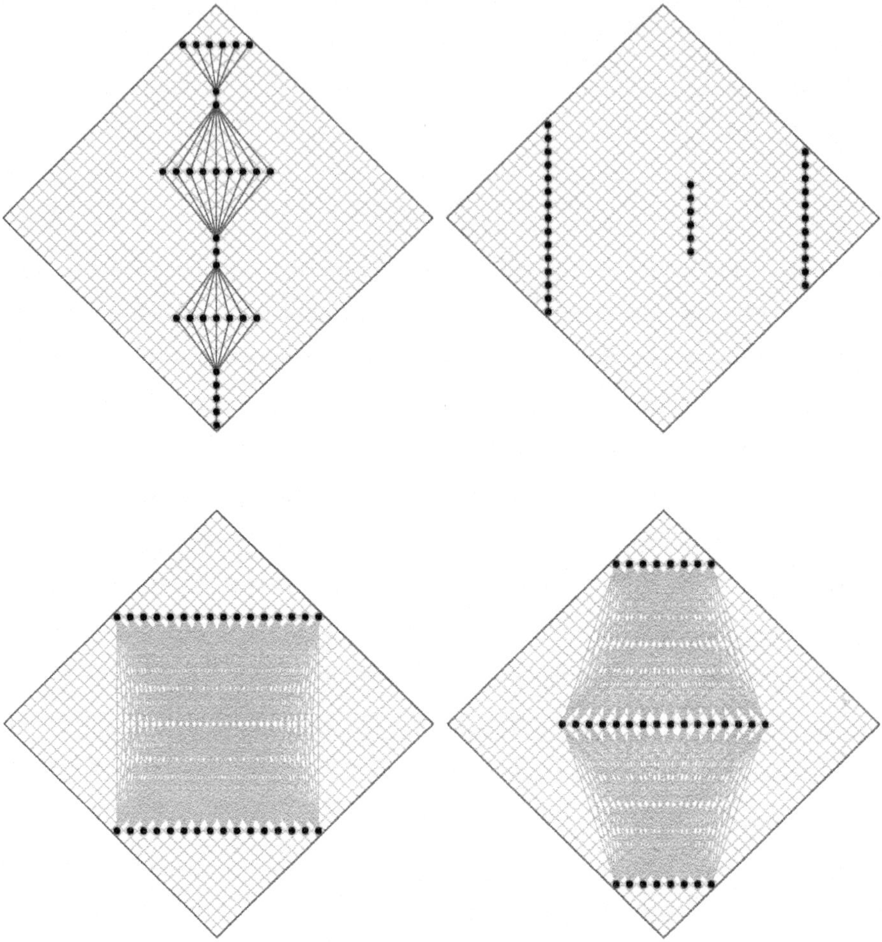

Fig. A.9 Examples of 2d orders of size $n = 32$ which embed into the causal diamond but not faithfully. They are not continuumlike. The two causal sets at the bottom are layered and maximally linked, and do not therefore represent the dominant bilayer posets or the KR posets of [20,21,50]

References

1. E.H. Kronheimer, R. Penrose, On the structure of causal spaces. Proc. Camb. Phil. Soc. **63**, 481–501 (1967). https://doi.org/10.1017/S030500410004144X
2. L. Bombelli, J. Lee, D. Meyer, R. Sorkin, Space-time as a causal set. Phys. Rev. Lett. **59**, 521–524 (1987). https://doi.org/10.1103/PhysRevLett.59.521
3. R. Loll, Quantum gravity from causal dynamical triangulations: a review. Class. Quant. Grav. **37**(1), 013002 (2020). https://doi.org/10.1088/1361-6382/ab57c7. (arXiv:1905.08669 [hep-th])
4. E. Bianchi, P. Martin-Dussaud, Causal structure in spin-foams, arXiv:2109.00986 [gr-qc]
5. B. Dittrich, J. Padua-Argüelles, Lorentzian quantum cosmology from effective spin foams. Universe **10**(7), 296 (2024). https://doi.org/10.3390/universe10070296. (arXiv:2306.06012 [gr-qc])
6. J. Madore, An introduction to noncommutative geometry. Lect. Notes Phys. **543**, 231–273 (2000). https://doi.org/10.1007/3-540-46552-9_5
7. F. Dowker, J. Butterfield, Recovering General Relativity from a Planck scale discrete theory of quantum gravity, arXiv:2106.01297 [gr-qc]
8. A. Robb, *A Theory of Time and Space* (University Press, Cambridge, 1914)
9. A.D. Alexandrov, V.V. Ovchinnikova, Notes on the foundations of relativity theory. Vestnik Leningrad. Univ. 95–110 (1953)
10. A.D. Alexandrov, A contribution to chronogeometry. Can. J. Math. **19**, 1119–1128 (1967)
11. E.C. Zeeman, Causality implies the lorentz group. J. Math. Phys. **5**(4), 490–493 (1964). https://doi.org/10.1063/1.1704140
12. S.W. Hawking, Particle creation by black holes. Commun. Math. Phys. **43**, 199–220 (1975). [Erratum: Commun.Math.Phys. 46, 206 (1976)]. https://doi.org/10.1007/BF02345020
13. S.W. Hawking, A.R. King, P.J. Mccarthy, A new topology for curved space-time which incorporates the causal, differential, and conformal structures. J. Math. Phys. **17**, 174–181 (1976). https://doi.org/10.1063/1.522874
14. D.B. Malament, The class of continuous timelike curves determines the topology of spacetime. J. Math. Phys. **18**(7), 1399–1404 (1977). https://doi.org/10.1063/1.523436
15. D. Finkelstein, Space-time code. Phys. Rev. **184**, 1261–1271 (1969). https://doi.org/10.1103/PhysRev.184.1261
16. A. Ashtekar, E. Bianchi, A short review of loop quantum gravity. Rept. Prog. Phys. **84**(4), 042001 (2021). https://doi.org/10.1088/1361-6633/abed91. (arXiv:2104.04394 [gr-qc])

S. Surya, *The Causal Set Approach to Quantum Gravity*, Lecture Notes in Physics 1036, https://doi.org/10.1007/978-3-031-84420-1

17. J. Polchinski, *String Theory, Cambridge Monographs on Mathematical Physics*, vol. 1 (Cambridge University Press, 1998). https://doi.org/10.1017/CBO9780511816079
18. J. Myrheim, Statistical geometry, Tech. Rep. CERN-TH-2538, CERN (1978)
19. M. Cortês, L. Smolin, Quantum energetic causal sets. Phys. Rev. D **90**(4), 044035 (2014). https://doi.org/10.1103/PhysRevD.90.044035. (arXiv:1308.2206 [gr-qc])
20. D.J. Kleitman, B.L. Rothschild, Asymptotic enumeration of partial orders on a finite set. Trans. Am. Math. Soc. **205**, 205–220 (1975)
21. D. Dhar, Entropy and phase transitions in partially ordered sets. J. Math. Phys. **19**(8), 1711–1713 (1978)
22. D. Dhar, Asymptotic enumeration of partially ordered sets. Pac. J. Math. **90**(2), 299 (1980)
23. F. Dowker, R.D. Sorkin, An intrinsic causality principle in histories-based quantum theory: a proposal. J. Phys. A **56**(48), 485304 (2023). https://doi.org/10.1088/1751-8121/ad0347. (arXiv:2305.16828 [quant-ph])
24. C. Wüthrich, The philosophy of causal set theory (2023)
25. J. Bell, *Quantum Gravity 2* (Clarendon Press, Oxford, 1981), pp.611–637
26. R.D. Sorkin, Quantum dynamics without the wave function. J. Phys. **A40**, 3207–3222 (2007). https://doi.org/10.1088/1751-8113/40/12/S20, (arXiv:quant-ph/0610204 [quant-ph])
27. R.D. Sorkin, Quantum mechanics as quantum measure theory. Mod. Phys. Lett. **A9**, 3119–3128 (1994). https://doi.org/10.1142/S021773239400294X, (arXiv:gr-qc/9401003 [gr-qc])
28. R.D. Sorkin, Quantum measure theory and its interpretation, in *Physics and experiments with linear colliders. Proceedings, 3rd Workshop, Morioka-Appi, Japan, September 8-12, 1995. Vol. 1, 2* (1995). arXiv:gr-qc/9507057 [gr-qc]
29. R.B. Salgado, Some identities for the quantum measure and its generalizations, Mod. Phys. Lett. **A17**, 711–728 (2002). https://doi.org/10.1142/S0217732302007041, (arXiv:gr-qc/9903015 [gr-qc])
30. R.D. Sorkin, An Exercise in "anhomomorphic logic.âŁž J. Phys. Conf. Ser. **67**, 012018 (2007). https://doi.org/10.1088/1742-6596/67/1/012018. (arXiv:quant-ph/0703276 [quant-ph])
31. F. Dowker, Y. Ghazi-Tabatabai, The Kochen-Specker theorem revisited in quantum measure theory. J. Phys. A **41**, 105301 (2008). https://doi.org/10.1088/1751-8113/41/10/105301. (arXiv:0711.0894 [quant-ph])
32. S. Chakraborti, R.D. Sorkin, U. Sinha, Measuring a "Probability" >1," arXiv:2407.15702 [quant-ph]
33. D. Meyer, *The Dimension of Causal Sets*. Ph.D. thesis, M.I.T. (1988)
34. S. Major, D. Rideout, S. Surya, On Recovering continuum topology from a causal set. J. Math. Phys. **48**, 032501 (2007). https://doi.org/10.1063/1.2435599. (arXiv:gr-qc/0604124 [gr-qc])
35. G. Brightwell, R. Gregory, The Structure of random discrete space-time. Phys. Rev. Lett. **66**, 260–263 (1991). https://doi.org/10.1103/PhysRevLett.66.260
36. D.M. Benincasa, F. Dowker, The scalar curvature of a causal set. Phys. Rev. Lett. **104**, 181301 (2010). https://doi.org/10.1103/PhysRevLett.104.181301
37. F. Dowker, L. Glaser, Causal set d'alembertians for various dimensions. Class. Quantum Gravity **30**(19), 195016 (2013)
38. S. Johnston, Feynman propagator for a free scalar field on a causal set. Phys. Rev. Lett. **103**, 180401 (2009). https://doi.org/10.1103/PhysRevLett.103.180401. (arXiv:0909.0944 [hep-th])
39. S.P. Johnston, *Quantum Fields on Causal Sets*. Ph.D. thesis, Imperial Coll., London (2010). arXiv:1010.5514 [hep-th]
40. R.D. Sorkin, From green function to quantum field. Int. J. Geom. Meth. Mod. Phys. **14**(08), 1740007 (2017). https://doi.org/10.1142/S0219887817400072. (arXiv:1703.00610 [gr-qc])
41. S. Surya, N.X, Y.K. Yazdi, Studies on the SJ vacuum in de sitter spacetime. JHEP **07**, 009 (2019). https://doi.org/10.1007/JHEP07(2019)009, arXiv:1812.10228 [gr-qc]
42. R.D. Sorkin, Expressing entropy globally in terms of (4D) field-correlations. J. Phys. Conf. Ser. **484**, 012004 (2014). https://doi.org/10.1088/1742-6596/484/1/012004. (arXiv:1205.2953 [hep-th])

43. R.D. Sorkin, Y.K. Yazdi, Entanglement entropy in causal set theory. Class. Quant. Grav. **35**(7), 074004 (2018). https://doi.org/10.1088/1361-6382/aab06f. (arXiv:1611.10281 [hep-th])
44. S. Surya, N. X, Y.K. Yazdi, Entanglement entropy of causal set de Sitter horizons. Class. Quant. Grav. **38**(11), 115001 (2021). https://doi.org/10.1088/1361-6382/abf279, arXiv:2008.07697 [gr-qc]
45. E. Dable-Heath, C.J. Fewster, K. Rejzner, N. Woods, Algebraic classical and quantum field theory on causal sets. Phys. Rev. D **101**(6), 065013 (2020). https://doi.org/10.1103/PhysRevD.101.065013. (arXiv:1908.01973 [math-ph])
46. I. Jubb, Interacting Scalar Field Theory on Causal Sets, arXiv:2306.12484 [hep-th]
47. E. Albertini, F. Dowker, A. Nasiri, S. Zalel, In-in correlators and scattering amplitudes on a causal set. Phys. Rev. D **109**(10), 106014 (2024). https://doi.org/10.1103/PhysRevD.109.106014. (arXiv:2402.08555 [hep-th])
48. R.D. Sorkin, Forks in the road, on the way to quantum gravity. Int. J. Theor. Phys. **36**, 2759–2781 (1997). https://doi.org/10.1007/BF02435709, arXiv:gr-qc/9706002 [gr-qc]
49. W.G. Unruh, R.M. Wald, Time and the interpretation of canonical quantum gravity. Phys. Rev. D **40**, 2598 (1989). https://doi.org/10.1103/PhysRevD.40.2598
50. H.J. Prömel, A. Steger, A. Taraz, Phase transitions in the evolution of partial orders. J. Comb. Theory, Ser. A **94**, 230–275 (2001). https://doi.org/10.1006/jcta.2000.3135
51. J. Ambjorn, S. Jain, G. Thorleifsson, Baby universes in 2-d quantum gravity. Phys. Lett. B **307**, 34–39 (1993). https://doi.org/10.1016/0370-2693(93)90188-N. (arXiv:hep-th/9303149)
52. S.P. Loomis, S. Carlip, Suppression of non-manifold-like sets in the causal set path integral. Class. Quant. Grav. **35**(2), 024002 (2018). https://doi.org/10.1088/1361-6382/aa980b. (arXiv:1709.00064 [gr-qc])
53. A. Mathur, A.A. Singh, S. Surya, Entropy and the link action in the causal set path-sum. Class. Quant. Grav. **38**(4), 045017 (2021). https://doi.org/10.1088/1361-6382/abd300. (arXiv:2009.07623 [gr-qc])
54. P. Carlip, S. Carlip, S. Surya, Path integral suppression of badly behaved causal sets, arXiv:2209.00327 [gr-qc]
55. P. Carlip, S. Carlip, S. Surya, The einstein-hilbert action for entropically dominant causal sets. In Preparation
56. S. Surya, Evidence for the continuum in 2d causal set quantum gravity. Class. Quantum Grav. **29**(13), 132001 (2012). https://doi.org/10.1088/0264-9381/29/13/132001
57. J.B. Hartle, S.W. Hawking, Wave function of the universe. Phys. Rev. D **28**(12), 2960–2975 (1983). https://doi.org/10.1103/PhysRevD.28.2960
58. J. Henson, D. Rideout, R.D. Sorkin, S. Surya, Onset of the asymptotic regime for (uniformly random) finite orders. Exp. Math. **26**(3), 253–266 (2017). https://doi.org/10.1080/10586458.2016.1158134
59. L. Glaser, D. O'Connor, S. Surya, Finite size scaling in 2d causal set quantum gravity. Class. Quant. Grav. **35**(4), 045006 (2018). https://doi.org/10.1088/1361-6382/aa9540. (arXiv:1706.06432 [gr-qc])
60. L. Glaser, The Ising model coupled to 2d orders. Class. Quant. Grav. **35**(8), 084001 (2018). https://doi.org/10.1088/1361-6382/aab139. (arXiv:1802.02519 [gr-qc])
61. W.J. Cunningham, S. Surya, Dimensionally restricted causal set quantum gravity: examples in two and three dimensions. Class. Quant. Grav. **37**(5), 054002 (2020). (arXiv:1908.11647 [gr-qc])
62. D.P. Rideout, R.D. Sorkin, A Classical sequential growth dynamics for causal sets. Phys. Rev. **D61**, 024002 (2000). https://doi.org/10.1103/PhysRevD.61.024002, arXiv:gr-qc/9904062 [gr-qc]
63. D.P. Rideout, R.D. Sorkin, Evidence for a continuum limit in causal set dynamics. Phys. Rev. **D63**, 104011 (2001). https://doi.org/10.1103/PhysRevD.63.104011, arXiv:gr-qc/0003117 [gr-qc]
64. X. Martin, D. O'Connor, D.P. Rideout, R.D. Sorkin, On the 'renormalization' transformations induced by cycles of expansion and contraction in causal set cosmology. Phys. Rev. **D63**, 084026 (2001). https://doi.org/10.1103/PhysRevD.63.084026, arXiv:gr-qc/0009063 [gr-qc]

65. G. Brightwell, H.F. Dowker, R.S. Garcia, J. Henson, R.D. Sorkin, Observables in causal set cosmology. Phys. Rev. D **67**, 084031 (2003). https://doi.org/10.1103/PhysRevD.67.084031
66. D.P. Rideout, *Dynamics of causal sets*. Ph.D. thesis, Syracuse U. (2001). arXiv:gr-qc/0212064 [gr-qc]
67. X. Martin, D. O'Connor, R.D. Sorkin, The Random walk in generalized quantum theory. Phys. Rev. D **71**, 024029 (2005). https://doi.org/10.1103/PhysRevD.71.024029. (arXiv:gr-qc/0403085)
68. R.D. Sorkin, An exercise in "anhomomorphic logic". J. Phys.: Conf. Ser. **67**(1), 012018 (2007). https://doi.org/10.1088/1742-6596/67/1/012018
69. F. Dowker, S. Johnston, R.D. Sorkin, Hilbert spaces from path integrals. J. Phys. A **43**, 275302 (2010). https://doi.org/10.1088/1751-8113/43/27/275302. (arXiv:1002.0589 [quant-ph])
70. F. Dowker, S. Johnston, S. Surya, On extending the quantum measure. J. Phys. A **43**, 505305 (2010). https://doi.org/10.1088/1751-8113/43/50/505305. (arXiv:1007.2725 [gr-qc])
71. S. Surya, S. Zalel, A criterion for covariance in complex sequential growth models. Class. Quant. Grav. **37**(19), 195030 (2020). https://doi.org/10.1088/1361-6382/ab987f. (arXiv:2003.11311 [math-ph])
72. L. Bombelli, J. Henson, R.D. Sorkin, Discreteness without symmetry breaking: a theorem. Mod. Phys. Lett. A **24**, 2579–2587 (2009). https://doi.org/10.1142/S0217732309031958
73. D. Mattingly, Modern tests of lorentz invariance. Living Rev. Rel. **8**(1), 5 (2005), arXiv:gr-qc/0502097 [gr-qc]
74. R. Sorkin, A modified sum-over-histories for gravity," in *Highlights in gravitation and cosmology: Proceedings of the International Conference on Gravitation and Cosmology*, ed. by J.V.N. B.R. Iyer, Ajit Kembhavi, C.V. Vishveshwara. Goa, India, 14-19 December 1987 (1987)
75. R.D. Sorkin, Spacetime and causal sets," in *Relativity and Gravitation: Classical and Quantum*, ed. by J. D'Olivo, ILARG VII Conference, Cocoyocan, Mexico (World Scientific, Singapore, 1991, 1990), pp. 150–173
76. M. Ahmed, S. Dodelson, P.B. Greene, R. Sorkin, Everpresent lambda. Phys. Rev. D **69**, 103523 (2004). https://doi.org/10.1103/PhysRevD.69.103523
77. M. Ahmed, R. Sorkin, Everpresent lambda II. Structural stability. Phys. Rev. D **87**(6), 063515 (2013). https://doi.org/10.1103/PhysRevD.87.063515. (arXiv:1210.2589 [gr-qc])
78. N. Zwane, N. Afshordi, R.D. Sorkin, Cosmological tests of Everpresent Λ. Class. Quant. Grav. **35**(19), 194002 (2018). https://doi.org/10.1088/1361-6382/aadc36. (arXiv:1703.06265 [gr-qc])
79. S. Das, A. Nasiri, Y.K. Yazdi, Aspects of Everpresent Λ (I): A Fluctuating Cosmological Constant from Spacetime Discreteness arXiv:2304.03819 [gr-qc]
80. S. Das, A. Nasiri, Y.K. Yazdi, Aspects of Everpresent Λ (II): Cosmological Tests of Current Models, arXiv:2307.13743 [astro-ph.CO]
81. G. Hemion, A quantum theory of space and time. Found. Phys. **10**, 819–840 (1980). https://doi.org/10.1007/BF00708682
82. S. Hawking, G. Ellis, *Large Scale Structure of Spacetime* (Cambridge University Press, Cambridge, 1973)
83. R. Penrose, *Techniques of Differential Topology in Relativity* (Society for industrial & applied mathematics, PA, 1972)
84. K. Gödel, An example of a new type of cosmological solutions of Einstein's field equations of gravitation. Rev. Mod. Phys. **21**, 447–450 (1949). https://doi.org/10.1103/RevModPhys.21.447
85. E. Minguzzi, Lorentzian causality theory. Living Rev. Rel. **22**(1), 3 (2019). https://doi.org/10.1007/s41114-019-0019-x
86. Y. Choquet-Bruhat, recent results on the cauchy problem for gravitation and Yang-Mills fieLDS.," in *The Second Marcel Grossmann Meeting on the Recent Developments of General Relativity (In Honor of Albert Einstein)*. (1979) 7
87. Y. Choquet-Bruhat, Beginnings of the Cauchy problem, arXiv:1410.3490 [gr-qc]
88. A. Robb, *Geometry of Time And Space* (At The University Press, 1936)
89. J.L. Synge, ed., *Relativity: The General theory* (North-Holland (January 1, 1966), 1960)

90. J.A. Lester, The causal automorphism of de sitter and Einstein cylinder spacetimes. J. Math. Phys. **25**, 113–116 (1984)
91. O. Parrikar, S. Surya, Causal topology in future and past distinguishing spacetimes. Class. Quant. Grav. **28**, 155020 (2011). https://doi.org/10.1088/0264-9381/28/15/155020. (arXiv:1102.0936 [gr-qc])
92. N.H. Christ, R. Friedberg, T.D. Lee, Random lattice field theory: general formulation. Nucl. Phys. **B202**, 89 (1982). [,3(1982)]. https://doi.org/10.1016/0550-3213(82)90222-X
93. D. Stoyan, W. Kendall, J. Mecke, *Stochastic Geometry and Its Applications* (Wiley, 1995)
94. L. Bombelli, Statistical Lorentzian geometry and the closeness of Lorentzian manifolds. J. Math. Phys. **41**, 6944–6958 (2000). https://doi.org/10.1063/1.1288494, arXiv:gr-qc/0002053 [gr-qc]
95. M. Saravani, S. Aslanbeigi, On the causal set-continuum correspondence. Class. Quant. Grav. **31**(20), 205013 (2014). https://doi.org/10.1088/0264-9381/31/20/205013. (arXiv:1403.6429 [hep-th])
96. F. Dowker, J. Henson, R.D. Sorkin, Quantum gravity phenomenology, Lorentz invariance and discreteness. Mod. Phys. Lett. **A19**, 1829–1840 (2004). https://doi.org/10.1142/S0217732304015026, arXiv:gr-qc/0311055 [gr-qc]
97. S. Liberati, D. Mattingly, Lorentz breaking effective field theory models for matter and gravity: theory and observational constraints (2012)
98. L. Bombelli, D.A. Meyer, The origin of lorentzian geometry. Phys. Lett. A **141**, 226–228 (1989). https://doi.org/10.1016/0375-9601(89)90474-X
99. K. Martin, P. Panangaden, A Domain of spacetime intervals in general relativity. Commun. Math. Phys. **267**, 563–586 (2006). https://doi.org/10.1007/s00220-006-0066-5, (arXiv:gr-qc/0407094)
100. M. Braun, Spacetime reconstruction by order and number (2025). arXiv: 2507.01907 [gr-qc]
101. B.R. Greene, M.R. Plesser, Mirror manifolds: a brief review and progress report, in *2nd International Symposium on Particles, Strings and Cosmology (PASCOS 1991) Boston, Massachusetts, March 25-30, 1991* (1991), pp. 0648–666. arXiv:hep-th/9110014 [hep-th]
102. J. Noldus, A Lorentzian Lipschitz, Gromov-Hausdoff notion of distance. Class. Quant. Grav. **21**, 839–850 (2004). https://doi.org/10.1088/0264-9381/21/4/007, arXiv:gr-qc/0308074 [gr-qc]
103. J. Noldus, A new topology on the space of Lorentzian metrics on a fixed manifold. Class. Quant. Grav. **19**, 6075–6107 (2002). https://doi.org/10.1088/0264-9381/19/23/313. (arXiv:1104.1811 [gr-qc])
104. L. Bombelli, J. Noldus, The Moduli space of isometry classes of globally hyperbolic spacetimes. Class. Quant. Grav. **21**, 4429–4454 (2004). https://doi.org/10.1088/0264-9381/21/18/010, arXiv:gr-qc/0402049 [gr-qc]
105. L. Bombelli, J. Noldus, J. Tafoya, Lorentzian Manifolds and Causal Sets as Partially Ordered Measure Spaces, arXiv:1212.0601 [gr-qc]
106. B. Allen, A. Burtscher, Properties of the null distance and spacetime convergence. Int. Math. Res. Not. **2022**(10), 7729–7808 (2022). https://doi.org/10.1093/imrn/rnaa311, arXiv:1909.04483 [math.DG]
107. E. Minguzzi, S. Suhr, Lorentzian metric spaces and their Gromov-Hausdorff convergence, arXiv:2209.14384 [math.DG]
108. M. Kunzinger , C. Sämann, Lorentzian length spaces. Annals Global Anal. Geom. **54**, 399–447 (2018), arXiv:1711.08990 [math.DG]
109. M. Kunzinger, R. Steinbauer, Null distance and convergence of Lorentzian length spaces, arXiv:2106.05393 [math.DG]
110. B. Bollobás, G. Brightwell, Box-spaces and random partial orders. Trans. Am. Math.l Soc. **324**(1), 59–72 (1991). http://www.jstor.org/stable/2001495
111. D.D. Reid, The manifold dimension of a causal set: tests in conformally flat spacetimes. Phys. Rev. D **67**, 024034 (2003). https://doi.org/10.1103/PhysRevD.67.024034. (arXiv:gr-qc/0207103)
112. E. Bachmat, Discrete spacetime and its applications, arXiv:gr-qc/0702140 [GR-QC]

113. D.D. Reid, The manifold dimension of a causal set: tests in conformally flat spacetimes. Phys. Rev. **D67**, 024034 (2003). https://doi.org/10.1103/PhysRevD.67.024034, arXiv:gr-qc/0207103 [gr-qc]
114. J. Kambor, N. X, Manifold properties from causal sets using chains. Class. Quant. Grav. **38**(2), 025012 (2020). https://doi.org/10.1088/1361-6382/abc8c9, arXiv:2007.03835 [gr-qc]
115. M. Roy, D. Sinha, a S. Surya, Discrete geometry of a small causal diamond. Phys. Rev. **D87**(4), 044046 (2013). https://doi.org/10.1103/PhysRevD.87.044046, arXiv:1212.0631 [gr-qc]
116. L. Bombelli , B.B. Pilgrim, A tale of two actions: a variational principle for two-dimensional causal sets. Class. Quant. Grav. **38**(8,) 085014 (2021). https://doi.org/10.1088/1361-6382/abe6b7, arXiv:2010.12985 [gr-qc]
117. R.P. Dilworth, A decomposition theorem for partially ordered sets. Ann. Math. **51**(1), 161–166 (1950)
118. M. Buck, F. Dowker, I. Jubb, S. Surya, Boundary terms for causal sets. Class. Quant. Grav. **32**(20), 205004 (2015). https://doi.org/10.1088/0264-9381/32/20/205004. (arXiv:1502.05388 [gr-qc])
119. I. Jubb, *Causal structure and quantum gravity*. Ph.D. thesis, Imperial Coll., London (2017). https://doi.org/10.25560/56610
120. G.W. Gibbons, S.N. Solodukhin, The Geometry of small causal diamonds. Phys. Lett. **B649**, 317–324 (2007) . https://doi.org/10.1016/j.physletb.2007.03.068, arXiv:hep-th/0703098 [hep-th]
121. S. Khetrapal, S. Surya, Boundary term contribution to the volume of a small causal diamond. Class. Quant. Grav. **30**,) 065005 (2013). https://doi.org/10.1088/0264-9381/30/6/065005, arXiv:1212.0629 [gr-qc]
122. J.R. Munkres, *Elements of Algebraic Topology* (Addison-Wesley, 1984)
123. G.D. Rham, complexes à automorphismes et homéomorphie différentiable". Annales de l'institut Fourier **2**, 51–67 (1950)
124. A. Weil, sur les theoremes de de rham", *Commentarii Mathematici Helvetici* (1952) 119–145
125. S. Major, D. Rideout, S. Surya, Stable homology as an indicator of manifoldlikeness in causal set theory. Class. Quant. Grav. **26**, 175008 (2009). https://doi.org/10.1088/0264-9381/26/17/175008. (arXiv:0902.0434 [gr-qc])
126. R. Wald, *General Relativity* (University of Chicago press, USA, 1984)
127. A. Eichhorn, S. Mizera, S. Surya, Echoes of asymptotic silence in causal set quantum gravity. Class. Quant. Grav. **34**(16), 16LT01 (2017). https://doi.org/10.1088/1361-6382/aa7d1b, arXiv:1703.08454 [gr-qc]
128. A. Eichhorn, S. Surya, F. Versteegen, Induced Spatial Geometry from Causal Structure, arXiv:1809.06192 [gr-qc]
129. A. Eichhorn, S. Mizera, Spectral dimension in causal set quantum gravity. Class. Quant. Grav. **31**, 125007 (2014). https://doi.org/10.1088/0264-9381/31/12/125007. (arXiv:1311.2530 [gr-qc])
130. A. Eichhorn, S. Surya, F. Versteegen, Spectral dimension on spatial hypersurfaces in causal set quantum gravity. Class. Quant. Grav. **36**(23), 235013 (2019). https://doi.org/10.1088/1361-6382/ab47cd. (arXiv:1905.13498 [gr-qc])
131. A. Ravishankar, Towards a discrete induced spatial curvature in causal sets, Master's thesis, Indian Institute of Science Education and Research Pune (2023)
132. L. Glaser, S. Surya, Towards a definition of locality in a manifoldlike causal set. Phys. Rev. D **88**(12), 124026 (2013). https://doi.org/10.1103/PhysRevD.88.124026. (arXiv:1309.3403 [gr-qc])
133. L. Glaser, *Endeavours in Discrete Lorentzian Geometry: A thesis in five papers*. Ph.D. thesis, Copenhagen U. (2014). arXiv:1409.8144 [gr-qc]
134. R.D. Sorkin, Does locality fail at intermediate length-scales, in *Approaches to Quantum Gravity*, ed. by D. Oriti (Cambridge University Press, 2007 pp. 26–43. arXiv: gr-qc/0703099
135. A. Daughton, *The Recovery of Locality for Causal Sets and Related Topics*. Ph.D. thesis, Syracuse University (1993)

136. R. Salgado, *Toward a Quantum Dynamics for Causal Sets*. Ph.D. thesis, Syracuse University (2008)
137. D.M.T. Benincasa, *The action of a casual set*. Ph.D. thesis, Imperial College London, 2013
138. L. Glaser, A closed form expression for the causal set d'alembertian. Class. Quant. Grav. **31**(9), 095007 (2014)
139. A. Belenchia, D.M.T. Benincasa, F. Dowker, The continuum limit of a 4-dimensional causal set scalar d'Alembertian. Class. Quant. Grav. **33**(24), 245018 (2016). https://doi.org/10.1088/0264-9381/33/24/245018. (arXiv:1510.04656 [gr-qc])
140. D.M. Benincasa, F. Dowker, B. Schmitzer, The random discrete action for 2-dimensional spacetime. Class. Quant. Grav. **28**, 105018 (2011). https://doi.org/10.1088/0264-9381/28/10/105018
141. D.M.T. Benincasa, F. Dowker, B. Schmitzer, The random discrete action for 2-dimensional spacetime. Class. Quant. Grav. **28**, 105018 (2011). https://doi.org/10.1088/0264-9381/28/10/105018. (arXiv:1011.5191 [gr-qc])
142. F. Dowker, Boundary contributions in the causal set action. Gen. Rel. Grav. **38**(7), 075018 (2021). https://doi.org/10.1088/1361-6382/abc2fd. (arXiv:2007.13206 [gr-qc])
143. L. Machet, J. Wang, On the continuum limit of Benincasa-Dowker-Glaser causal set action. Class. Quant. Grav. **38**(1), 015010 (2021). https://doi.org/10.1088/1361-6382/abc274. (arXiv:2007.13192 [gr-qc])
144. D. Dou, R.D. Sorkin, Black hole entropy as causal links. Found. Phys. **33**, 279–296 (2003). https://doi.org/10.1023/A:1023781022519, arXiv:gr-qc/0302009 [gr-qc]
145. C. Barton, A. Counsell, F. Dowker, D.S.W. Gould, I. Jubb, G. Taylor, Horizon molecules in causal set theory. Phys. Rev. D **100**(12), 126008 (2019). https://doi.org/10.1103/PhysRevD.100.126008. (arXiv:1909.08620 [gr-qc])
146. D. Rideout, P. Wallden, Spacelike distance from discrete causal order. Class. Quant. Grav. **26**, 155013 (2009). https://doi.org/10.1088/0264-9381/26/15/155013. (arXiv:0810.1768 [gr-qc])
147. W. Cunningham, Inference of boundaries in causal sets. Class. Quant. Grav. **35**(9), 094002 (2018). https://doi.org/10.1088/1361-6382/aaadc4. (arXiv:1710.09705 [gr-qc])
148. S. Carlip, Dimensional reduction in causal set gravity. Class. Quant. Grav. **32**(23), 232001 (2015). https://doi.org/10.1088/0264-9381/32/23/232001. (arXiv:1506.08775 [gr-qc])
149. J. Henson, Constructing an interval of Minkowski space from a causal set. Class. Quant. Grav. **23**, L29–L35 (2006). https://doi.org/10.1088/0264-9381/23/4/L02, arXiv:gr-qc/0601069 [gr-qc]
150. Y.K. Yazdi, A. Kempf, Towards spectral geometry for causal sets. Class. Quant. Grav. **34**(9), 094001 (2017). https://doi.org/10.1088/1361-6382/aa663f. (arXiv:1611.09947 [hep-th])
151. Y.K. Yazdi, M. LetiziaA, Kempf, Lorentzian spectral geometry with causal sets. Class. Quant. Grav. **38**(1), 015011 (2021). https://doi.org/10.1088/1361-6382/abc3f8. (arXiv:2008.02291 [hep-th])
152. M. Aghili, L. Bombelli, B. Pilgrim, Discrete spacetime: a web of chains. Class. Quant. Grav. **36**(18), 185015 (2019). https://doi.org/10.1088/1361-6382/ab30e3. (arXiv:1807.08701 [gr-qc])
153. M.E. Aghili, L. Bombelli, B.B. Pilgrim, Path length distribution in two-dimensional causal sets. Eur. Phys. J. C **78**(9), 744 (2018). https://doi.org/10.1140/epjc/s10052-018-6229-7. (arXiv:1805.07312 [gr-qc])
154. H. Liu, D.D. Reid, A new approach for embedding causal sets into Minkowski space. Class. Quant. Grav. **35**(12), 124002 (2018). https://doi.org/10.1088/1361-6382/aac082
155. S. Johnston, Particle propagators on discrete spacetime. Class. Quant. Grav. **25**, 202001 (2008). https://doi.org/10.1088/0264-9381/25/20/202001. (arXiv:0806.3083 [hep-th])
156. N. Afshordi, S. Aslanbeigi, R.D. Sorkin, A distinguished vacuum state for a quantum field in a curved spacetime: formalism, features, and cosmology. JHEP **08**, 137 (2012). https://doi.org/10.1007/JHEP08(2012)137. (arXiv:1205.1296 [hep-th])
157. F. Dowker, S. Surya, N. X, Scalar field green functions on causal sets. Class. Quant. Grav. **34**(12,) 124002 (2017). https://doi.org/10.1088/1361-6382/aa6bc7, arXiv:1701.07212 [gr-qc]

158. N. X, *Aspects of Quantum Fields on Causal Sets*. Ph.D. thesis, Jawaharlal Nehru University (2021), arXiv:2105.07241 [gr-qc]
159. C.J. Fewster, R. Verch, On a recent construction of "Vacuum-likeâŁž quantum field states in curved spacetime. Class. Quant. Grav. **29**, 205017 (2012). https://doi.org/10.1088/0264-9381/29/20/205017. (arXiv:1206.1562 [math-ph])
160. M. Brum, K. Fredenhagen, 'Vacuum-like' Hadamard states for quantum fields on curved spacetimes. Class. Quant. Grav. **31**, 025024 (2014). https://doi.org/10.1088/0264-9381/31/2/025024. (arXiv:1307.0482 [gr-qc])
161. N. Afshordi, M. Buck, F. Dowker, D. Rideout, R.D. Sorkin, Y.K. Yazdi, A ground state for the causal diamond in 2 dimensions. JHEP **10**, 088 (2012). https://doi.org/10.1007/JHEP10(2012)088. (arXiv:1207.7101 [hep-th])
162. M. Saravani, R.D. Sorkin, Y.K. Yazdi, Spacetime entanglement entropy in 1 + 1 dimensions. Class. Quant. Grav. **31**(21), 214006 (2014). https://doi.org/10.1088/0264-9381/31/21/214006. (arXiv:1311.7146 [hep-th])
163. A. Belenchia, D.M.T. Benincasa, M. Letizia, S. Liberati, On the entanglement entropy of quantum fields in causal sets. Class. Quant. Grav. **35**(7), 074002 (2018). https://doi.org/10.1088/1361-6382/aaae27. (arXiv:1712.04227 [gr-qc])
164. Y.K. Yazdi, *Entanglement Entropy of Scalar Fields in Causal Set Theory*. Ph.D. thesis, Waterloo U. (2017)
165. N. X, Quantum Field Theory On Causal Sets, arXiv:2306.04800 [hep-th]
166. A. Bhattacharya, A. Mathur, S. Surya, Null geodesics from ladder molecules. Gen. Rel. Grav. **55**(2), 32 (2023). https://doi.org/10.1007/s10714-023-03074-y. (arXiv:2301.06480 [gr-qc])
167. T.S. Bunch, L. Parker, Feynman propagator in curved space-time: a momentum space representation. Phys. Rev. D **20**, 2499–2510 (1979). https://doi.org/10.1103/PhysRevD.20.2499
168. R.D. Sorkin, Scalar field theory on a causal set in histories form. J. Phys. Conf. Ser. **306**, 012017 (2011). https://doi.org/10.1088/1742-6596/306/1/012017. (arXiv:1107.0698 [gr-qc])
169. M. Brum, K. Fredenhagen, 'Vacuum-like' Hadamard states for quantum fields on curved spacetimes. Class. Quant. Grav. **31**, 025024 (2014). https://doi.org/10.1088/0264-9381/31/2/025024. (arXiv:1307.0482 [gr-qc])
170. C.J. Fewster, The art of the state. Int. J. Mod. Phys. D **27**(11), 1843007 (2018). https://doi.org/10.1142/S0218271818430071. (arXiv:1803.06836 [gr-qc])
171. A. Mathur, S. Surya, X. Nomaan, Spacetime entanglement entropy: covariance and discreteness. Gen. Rel. Grav. **54**(7), 74 (2022). https://doi.org/10.1007/s10714-022-02948-x. (arXiv:2207.01080 [gr-qc])
172. N. Afshordi, M. Buck, F. Dowker, D. Rideout, R.D. Sorkin, Y.K. Yazdi, A ground state for the causal diamond in 2 dimensions. JHEP **10**, 088 (2012). https://doi.org/10.1007/JHEP10(2012)088. (arXiv:1207.7101 [hep-th])
173. A. Mathur, S. Surya, Sorkin-Johnston vacuum for a massive scalar field in the 2D causal diamond. Phys. Rev. D **100**(4), 045007 (2019). https://doi.org/10.1103/PhysRevD.104.089901, arXiv:1906.07952 [hep-th]. [Erratum: Phys.Rev.D 104, 089901 (2021)]
174. S. Surya, N.X, Y.K. Yazdi, Studies on the SJ Vacuum in de Sitter Spacetime, arXiv:1812.10228 [gr-qc]
175. W.G. Unruh, Notes on black hole evaporation. Phys. Rev. D **14**, 870 (1976). https://doi.org/10.1103/PhysRevD.14.870
176. G.W. Gibbons, S.W. Hawking, Action integrals and partition functions in quantum gravity. Phys. Rev. D **15**, 2752–2756 (1977). https://doi.org/10.1103/PhysRevD.15.2752
177. L. Bombelli, R.K. Koul, J. Lee, R.D. Sorkin, A quantum source of entropy for black holes. Phys. Rev. D **34**, 373–383 (1986). https://doi.org/10.1103/PhysRevD.34.373
178. D. Vodola, L. Lepori, E. Ercolessi, G. Pupillo, Long-range ising and kitaev models: phases, correlations and edge modes. New J. Phys. **18**(1), 015001 (2015)
179. B. Basa, G. La Nave, P.W. Phillips, Classification of nonlocal actions: area versus volume entanglement entropy. Phys. Rev. D **101**(10), 106006 (2020). https://doi.org/10.1103/PhysRevD.101.106006. (arXiv:1907.09494 [hep-th])
180. J. Dougherty, C. Callender, Black hole thermodynamics: More than an analogy? (2016)

181. M. Saravani, S. Aslanbeigi, Dark matter from spacetime nonlocality. Phys. Rev. D **92**(10), (2015). https://doi.org/10.1103/PhysRevD.92.103504, arXiv:1502.01655 [hep-th]
182. M. Saravani, N. Afshordi, Off-shell dark matter: a cosmological relic of quantum gravity. Phys. Rev. D **95**(4), 043514 (2017). https://doi.org/10.1103/PhysRevD.95.043514
183. G.P. de Brito, A. Eichhorn, L. Fausten, Towards a bound on the Higgs mass in causal set quantum gravity, arXiv:2305.07595 [gr-qc]
184. L. Glaser, S. Surya, The Hartle-Hawking wave function in 2D causal set quantum gravity. Class. Quant. Grav. **33**(6), 065003 (2016). https://doi.org/10.1088/0264-9381/33/6/065003. (arXiv:1410.8775 [gr-qc])
185. A. Dhingra, L. Glaser, S. Surya, Work in progress
186. G.P. de Brito, A. Eichhorn, C. Pfeiffer, Higher-order curvature operators in causal set quantum gravity. Eur. Phys. J. Plus **138**(7), 592 (2023). https://doi.org/10.1140/epjp/s13360-023-04202-y. (arXiv:2301.13525 [gr-qc])
187. F. Dowker, Y. Ghazi-Tabatabai, The Kochen-Specker theorem revisited in quantum measure theory. J. Phys. A **41**, 105301 (2008). https://doi.org/10.1088/1751-8113/41/10/105301. (arXiv:0711.0894 [quant-ph])
188. F. Dowker, R.D. Sorkin, An intrinsic causality principle in histories-based quantum theory: a proposal. J. Phys. A **56**(48), 485304 (2023). https://doi.org/10.1088/1751-8121/ad0347. (arXiv:2305.16828 [quant-ph])
189. H.J. Prömel, A. Steger, A. Taraz, Phase transitions in the evolution of partial orders. J. Combin. Theory, Ser. A **94**, 230–275 (2001). https://doi.org/10.1006/jcta.2000.3135
190. M.E.J. Newman, G.T. Barkema, *Monte Carlo Methods in Statistical Physics* (Clarendon Press, Feb., 1999)
191. H. J. Prömel, Counting unlabeled structures. J. Comb. Theory A **44**, 83–93 (1987). https://api.semanticscholar.org/CorpusID:27988852
192. H.J. Promel, A. Steger, A. Taraz, Counting partial orders with a fixed number of comparable pairs. Combin. Prob. Comput. **10**(2), 159–177 (2001). https://doi.org/10.1017/S0963548301004503
193. P. Winkler, Random orders of dimension 2. Order **7**, 329 (1991)
194. M. El-Zahar, N. Sauer, Counting two-dimensional posets. Order **5**, 239 (1988)
195. R.F. Streater, A.S. Wightman, *PCT, Spin and Statistics, and All That* (Princeton University Press, 1989)
196. M. Varadarajan, D. Rideout, A general solution for classical sequential growth dynamics of causal sets. Phys. Rev. **D73**, 104021 (2006). https://doi.org/10.1103/PhysRevD.73.104021, arXiv:gr-qc/0504066 [gr-qc]
197. F. Dowker, S. Surya, Observables in extended percolation models of causal set cosmology. Class. Quant. Grav. **23**, 1381–1390 (2006). https://doi.org/10.1088/0264-9381/23/4/018, arXiv:gr-qc/0504069 [gr-qc]
198. F. Dowker, S. Zalel, Evolution of universes in causal set cosmology. Comptes Rendus Physique **18**, 246–253 (2017). https://doi.org/10.1016/j.crhy.2017.03.002. (arXiv:1703.07556 [gr-qc])
199. F. Dowker, N. Imambaccus, A. Owens, R. Sorkin, S. Zalel, A manifestly covariant framework for causal set dynamics. Class. Quant. Grav. **37**(8), 085003 (2020). https://doi.org/10.1088/1361-6382/ab719c. (arXiv:1910.07292 [gr-qc])
200. M. Ahmed, D. Rideout, Indications of de Sitter spacetime from classical sequential growth dynamics of causal sets. Phys. Rev. D **81**, 083528 (2010). https://doi.org/10.1103/PhysRevD.81.083528. (arXiv:0909.4771 [gr-qc])
201. S. Zalel, *Discrete Random Spacetimes: Covariance and Quantization in Growth Dynamics for Causal Sets*. Ph.D. thesis, Imperial Coll., London (2021). https://doi.org/10.25560/93625
202. F. Dowker, The birth of spacetime atoms as the passage of time," in *Do We Need a Physics of 'Passage'? Cape Town, South Africa, December 10-14, 2012* (2014). https://doi.org/10.1111/nyas.12542, arXiv:1405.3492 [gr-qc]
203. N. Huggett, Skeptical notes on a physics of passage. Ann. N.Y. Acad. Sci. **1326**, 9 (2014)
204. C. Wuthrich, C. Callender, What becomes of a causal set?. Brit. J. Phil. Sci. **68**(3), 907–925 (2017). https://doi.org/10.1093/bjps/axv040, arXiv:1502.00018 [physics.hist-ph]

205. J.S. Bell, A. Aspect, *Speakable and Unspeakable in Quantum Mechanics: Collected Papers on Quantum Philosophy*, 2nd edn (Cambridge University Press, 2004)
206. C. Wuthrich, C. Callender, What becomes of a causal set?. Brit. J. Phil. Sci. **68**(3), 907–925 (2017). https://doi.org/10.1093/bjps/axv040, arXiv:1502.00018 [physics.hist-ph]
207. B.V. Bento, F. Dowker, S. Zalel, If time had no beginning: growth dynamics for past-infinite causal sets. Class. Quant. Grav. **39**(4), 045002 (2022). https://doi.org/10.1088/1361-6382/ac445f. (arXiv:2109.10749 [gr-qc])
208. G. Brightwell, N. Georgiou, Continuum limits for classical sequential growth models. Rand. Struct. Alg. **36**, 218–250 (2010)
209. F. Dowker, S. Surya, Observables in extendcarliped percolation models of causal set cosmology. Class. Quant. Grav. **23**, 1381–1390 (2006). https://doi.org/10.1088/0264-9381/23/4/018, arXiv:gr-qc/0504069 [gr-qc]
210. P.R. Halmos, *Measure Theory, Graduate Texts in Mathematics*, vol. 18 (Springer, New York, 1976)
211. R.P. Geroch, E.H. Kronheimer, R. Penrose, Ideal points in space-time. Proc. Roy. Soc. Lond. A **327**, 545–567 (1972). https://doi.org/10.1098/rspa.1972.0062
212. S. Zalel, Covariant Growth Dynamics, arXiv:2302.10582 [gr-qc]
213. F. Dowker, S. Zalel, Observables for cyclic causal set cosmologies. Class. Quant. Grav. **40**(15), 155015 (2023). https://doi.org/10.1088/1361-6382/ace149. (arXiv:2212.01149 [gr-qc])
214. J.J. Diestel, J.J. Uhl, *Vector Measures, Mathematical Surveys and Monographs*, vol. 15 (American Mathematical Society, 1977)
215. J. Henson, Causality, Bell's theorem, and Ontic Definiteness, arXiv:1102.2855 [quant-ph]
216. W. Rudin, *Real and Complex Analysis* (McGraw-Hill, New York, 1987)
217. G. Brightwell, M. Luczak, The mathematics of causal sets, arXiv:1510.05612 [math.CO]
218. S. Perlmutter, G. Aldering, G. Goldhaber, R.A. Knop, P. Nugent, P.G. Castro, S. Deustua, S. Fabbro, A. Goobar, D.E. Groom, I.M. Hook, A.G. Kim, M.Y. Kim, J.C. Lee, N.J. Nunes, R. Pain, C.R. Pennypacker, R. Quimby, C. Lidman, R.S. Ellis, M. Irwin, R.G. McMahon, P. Ruiz-Lapuente, N. Walton, B. Schaefer, B.J. Boyle, A.V. Filippenko, T. Matheson, A.S. Fruchter, N. Panagia, H.J.M. Newberg, W.J. Couch, T.S.C. Project, measurements of omega lambda from 42 high-redshift supernovae. APJ **517**(2), 565–586 (1999), arXiv:astro-ph/9812133 [astro-ph]
219. A.G. Riess, A.V. Filippenko, P. Challis, A. Clocchiatti, A. Diercks, P.M. Garnavich, R.L. Gilliland, C.J. Hogan, S. Jha, R.P. Kirshner, B. Leibundgut, M.M. Phillips, D. Reiss, B.P. Schmidt, R.A. Schommer, R.C. Smith, J. Spyromilio, C. Stubbs, N.B. Suntzeff, J. Tonry, Observational evidence from supernovae for an accelerating universe and a cosmological constant. Astron. J. **116**(3), 1009–1038 (1998). https://doi.org/10.1086/300499
220. B.P. Schmidt, N.B. Suntzeff, M.M. Phillips, R.A. Schommer, A. Clocchiatti, R.P. Kirshner, P. Garnavich, P. Challis, B. Leibundgut, J. Spyromilio, A.G. Riess, A.V. Filippenko, M. Hamuy, R.C. Smith, C. Hogan, C. Stubbs, A. Diercks, D. Reiss, R. Gilliland, J. Tonry, J. Maza, A. Dressler, J. Walsh, R. Ciardullo, The high-z supernova search: measuring cosmic deceleration and global curvature of the universe using type ia supernovae. Astrophys. J. **507**(1), 46–63 (1998). https://doi.org/10.1086/306308
221. L. Philpott, F. Dowker, R.D. Sorkin, Energy-momentum diffusion from spacetime discreteness. Phys. Rev. D **79**, 124047 (2009). https://doi.org/10.1103/PhysRevD.79.124047. (arXiv:0810.5591 [gr-qc])
222. C.R. Contaldi, F. Dowker, L. Philpott, Polarization diffusion from spacetime uncertainty. Class. Quant. Grav. **27**, 172001 (2010). https://doi.org/10.1088/0264-9381/27/17/172001. (arXiv:1001.4545 [gr-qc])
223. N. Kaloper, D. Mattingly, Low energy bounds on Poincare violation in causal set theory. Phys. Rev. D **74**, 106001 (2006). https://doi.org/10.1103/PhysRevD.74.106001. (arXiv:astro-ph/0607485)
224. E. Albertini, A. Nasiri, E. Panella, Stochastic Dark Matter: Covariant Brownian Motion from Planckian Discreteness, arXiv:2409.02188 [gr-qc]

225. R.D. Sorkin, Big extra dimensions make lambda too small. Braz. J. Phys. **35**, 280–283 (2005). https://doi.org/10.1590/S0103-97332005000200012, arXiv:gr-qc/0503057 [gr-qc]
226. J. Samuel, S. Sinha, Surface tension and the cosmological constant. Phys. Rev. Lett. **97**, 161302 (2006). https://doi.org/10.1103/PhysRevLett.97.161302, arXiv:cond-mat/0603804 [cond-mat]
227. J.D. Barrow, A strong constraint on ever-present lambda. Phys. Rev. **D75**, 067301 (2007). https://doi.org/10.1103/PhysRevD.75.067301, arXiv:gr-qc/0612128 [gr-qc]
228. J.A. Zuntz, The cosmic microwave background in a causal set universe. Phys. Rev. D **77**, 043002 (2008). https://doi.org/10.1103/PhysRevD.77.043002. (arXiv:0711.2904 [astro-ph])
229. E. Albertini, A. Nasiri, E. Panella, Stochastic Dark Matter: Covariant Brownian Motion from Planckian Discreteness, arXiv:2409.02188 [gr-qc]
230. R.D. Sorkin, Stochastic evolution on a manifold of states. Annals Phys. **168**, 119–147 (1986). https://doi.org/10.1016/0003-4916(86)90113-2
231. A. Belenchia, D.M.T. Benincasa, S. Liberati, E. Martin-Martinez, Transmission of information in nonlocal field theories. Phys. Rev. D **96**(11), 116006 (2017). https://doi.org/10.1103/PhysRevD.96.116006. (arXiv:1707.01654 [quant-ph])
232. A. Belenchia, D.M.T. Benincasa, S. Liberati, F. Marin, F. Marino, A. Ortolan, Tests of quantum-gravity-induced nonlocality via optomechanical experiments. Phys. Rev. D **95**(2), 026012 (2017). https://doi.org/10.1103/PhysRevD.95.026012. (arXiv:1611.07959 [gr-qc])
233. A. Belenchia, D.M.T. Benincasa, E. Martin-Martinez, M. Saravani, Low energy signatures of nonlocal field theories. Phys. Rev. **D94**(6), 061902(2016). https://doi.org/10.1103/PhysRevD.94.061902, arXiv:1605.03973 [quant-ph]
234. A. Belenchia, D.M.T. Benincasa, S. Liberati, F. Marin, F. Marino, A. Ortolan, Testing quantum gravity induced nonlocality via optomechanical quantum oscillators. Phys. Rev. Lett. **116**(16), 161303 (2016). https://doi.org/10.1103/PhysRevLett.116.161303. (arXiv:1512.02083 [gr-qc])
235. A. Belenchia, D.M.T. Benincasa, S. Liberati, Nonlocal scalar quantum field theory from causal sets. JHEP **03**, 036 (2015). https://doi.org/10.1007/JHEP03(2015)036. (arXiv:1411.6513 [gr-qc])
236. F. Dowker, J. Henson, R. Sorkin, Discreteness and the transmission of light from distant sources. Phys. Rev. D **82**, 104048 (2010). https://doi.org/10.1103/PhysRevD.82.104048. (arXiv:1009.3058 [gr-qc])
237. M. Saravani, *Aspects of Nonlocality: from Particles to Black Holes*. Ph.D. thesis, Waterloo U. (2016)
238. G. Brinkmann, B. McKay, Posets on up to 16 points. Order **19**, 147–179 (2002)
239. G. Brinkmann, B.D. McKay, Number of partially ordered sets "posets") with n unlabeled elements.. The on-line encyclopedia of integer sequences, published electronically at https://oeis.org,. https://oeis.org/A000112. Sequence A000112
240. H. Jamke, Number of partially ordered sets ("posets") with n labeled elements (or labeled acyclic transitive digraphs).. The on-line encyclopedia of integer sequences, published electronically at https://oeis.org,. https://oeis.org/A001035. Sequence A001035

Index

S. Surya, *The Causal Set Approach to Quantum Gravity*, Lecture Notes in Physics 1036, https://doi.org/10.1007/978-3-031-84420-1

The manufacturer's authorised representative in the EU is Springer Nature Customer Service Centre GmbH, Europaplatz 3, 69115 Heidelberg, Germany. If you have any concerns regarding our products, please contact ProductSafety@springernature.com

Printed and bound by CPI Group (UK) Ltd, Croydon, CR0 4YY

31/07/2026

02182787-0003